Best · Restaurant

KUNSAN DESIGN

@ Copyright 2025 **KUNSAN DESIGN** Co,Ltd

Publish	kunsan design
Publisher	Kim sun nam
Editorial Derector	Kim young sik
Writer	Kim in chul
Address	440-14, Buil-ro, Wonmi-gu, Bucheon-si, Gyeonggi-do, Republic of Korea
Telephone	031-547-2568
Fax	031-547-2569
Website	http://www.kunsan.com
E-mail	kunsandesign@gmail.com

이책에 게재된 내용 및 사진의 허락없는 무단 복제 및 전재를 금지합니다.
KUNSAN DESIGN publishing 2025 , All rights reserved.

Restaurant interior & Branding design?

When eating out, a good review isn't solely down to the food; a restaurant interior plays an important role in the whole experience. Knowing how to create an environment that complements the menu and the space's interior architecture is no simple feat.

7 things to consider when designing a restaurant.

1. Choose a Striking Colour Palette.

As with domestic projects, high up on the subject list for restaurant decor ideas is that of the colour scheme. "It's quite common for restaurant to be bold with their colour choices. In a sitting room or dining room there are just a few items of furniture and the colours chosen feel very much 'on show', in a restaurant setting, the volumes of furniture, accessories and light sources are far greater so there's a distraction from the colour. It's this that encourages the bravery and exploration in colour," Colour in a restaurant can also become a huge talking point. In larger establishments, it acts as a tool to define distinct areas. The colour becomes an icon for the bar in question. Remember, also, with colour decisions to reflect on how they sit together in daylight and how they evolve when night falls and they rely on candle and lamp light for luminosity.

2. Master a Functional Layout.

"A restaurant floor plan forms an immediate impression of ambiance. Keep tables spread out so that each feels secluded, and the guest experience will be an intimate one. "Tables packed tightly together on the other hand is a statement of conviviality and liveliness. Determining restaurant layout requires a decision on what atmosphere is hoped to be established,' There's a purely functional aspect to the restaurant room plan too though. The traffic should flow seamlessly. Bottlenecks are immediately apparent and cause an awkward distraction that guests will pick up on. A division between the restaurant area is another question to be answered on the subject of layout. This will affect the feeling of formality or informality across the whole space.

3. Specify High-Grade Contract Furniture.

When a bar approaches its furniture choices as objects of function and nothing else, the entire experience becomes devalued. The importance of carefully selected bar chairs, tables and accent pieces is not to be underestimated. "It matters that not every element matches. That doesn't mean that every dining table and chair needs to be different, but that there are occasional pieces in the room to break up any consistency. A statement dresser or several elegant console tables adorned with decorative lamps or a vase are helpful here and serve as a reminder that restaurant furniture extends beyond table and chair."

4. Select Show-stopping Lighting.

Arguably one of the most crucial aspects of restaurant design, lighting ideas must be respectful of the fine line between necessary, task-style beams and ambient illumination. "In our homes, we consistently stress the need to layer the lighting throughout the space. It's no good having all of the lighting hung from above; it must drift down slowly from pendants to wall lights and lamps aplenty. A restaurant may be a commercial space, but guests want to be made to feel at ease in their surroundings as they do at home, so it's logical to follow the same lighting philosophy as you would in the home." Also acknowledges the effect that lighting has on each guest, considering what is the most flattering light at every angle. In the Harrods Dining Hall, they focus the lighting on the plate to showcase the food, with soft, low-level lighting for the diners to bask in.

5. Curate a Unique Decor Collection.

"If there's exquisite detail in the food, then there should be exquisite detail in the way the room is put together." Restaurant decor ideas, therefore, are a fundamental component in how the approach every commercial project. "The term 'finishing touches' is misleading. These aren't the bits to be simply added in at the end. Accessories and decorative touches are where you deepen the level of consideration in restaurant interior design," Remember too that in the social media-centric world that we live in, carving out areas of appealing vignettes, complete with trinket boxes, decorative bowls and impressive floral arrangements, means your restaurant design is all the more likely to become a must-visit and must-photograph location.

6. Display Personality-Full Artwork.

By extension of decor and accessories, displaying expertly curated arts and artefacts speaks volumes to guests. They serve as points of interest, they reinforce or subvert the overall interior design direction and they add warmth and texture to the space. "Similar to how we light a restaurant with as much thought as we do a home's dining area, bar wall design deserves as much thought as would be given in a residential project. Art adorning the walls and sculptures atop of tables are luxurious details that reveal the pedigree of the establishment. We always encourage this in our restaurant interior design projects, and it's an investment that our clients never regret."

7. Don't Neglect The restaurant Bathroom Design.

It's not uncommon to hear creative types and interiors aficionados claim that to know the true dedication to design of a restaurant, you must check out the bathroom. This is where the same level of attention to detail seen in the main restaurant either flails or flourishes. "The combination and contrast of materiality is so interesting in bathrooms. Marble and stone give a luxurious finish and contrast beautifully with timber, which can add warmth. Mirrored glass, brass and nickel all give reflection and a sense of luminosity." Lighting, too, is fundamental in this area of the bar, otherwise undoing all of the hard work in setting the perfect level of ambiance during dining.

Table of contents

008~025 **Sushi Kaiyo** Brandworks

026~047 **Tero Brussels** Creneau International

048~057 **Sushi & Co.** Bond agency

058~093 **Osteria Betulla** DA bureau

094~107 **Japonesque Chef Restaurant** Baranowitz & Goldberg Architects, Pitsou Kedem Architects

108~119 **Jufka restaurant** Maden Group

120~133 **Hokkaidon** A work of substance

134~147 **Grain restaurant** DA bureau

148~161 **Tre Dining** InSpace Creative

162~181 **Pukkel Restaurant** Masquespacio

182~205 **Mama Manana** Masquespacio

206~223 **Green Grass Condesa** Taller David Dana

224~243 **Yilong Dim Sum Society** BrandWorks

244~257 **1111 Ones** M.R. Studio

258~277 **Fujiwara Yoshi** Sergey Makhno Architects

278~295 **Margo restaurant** M.R. Studio

296~303 **Restaurant Lunar** So Studio

▶Sushi Kaiyo: Daerah Khusus Ibukota Jakarta 11610, Indonesia / +62 877-2526-2721

Jakarta, Indonesia
Sushi Kaiyo
Crafted by the forces of nature — Japanese all-day dining in Jakarta.

Indonesia, and in particular Jakarta, has a strong appreciation for Japanese cuisine and culture. The category is a proven cuisine for Food & Beverage (F&B) due to the high demand for its style of dining. As a result, in Jakarta alone there are many Japanese-themed restaurants, grab 'n go and drinking bars. Out of the high number of competitors in this market, there are two F&B groups, in particular, that stand out from the crowd.

Founded in 2002, Boga Group is considered one of Indonesia's leading food and hospitality groups with over 150 restaurants spread all throughout the nation. Respected brands such as Bakerzin, Pepper Lunch, Paradise Dynasty are just to name a few. Adding to this landscape is a new concept soon to launch as part of the growing portfolio. In early 2020, the founder and CEO of Boga Group Indonesia engaged BrandWorks to create a new Japanese dining concept that would bring a fresh, innovative and exciting perspective to an already proven dining category.

As a key player in the F&B landscape with a strong reputation in the industry, Boga Group has operated various Japanese food concepts with significant resources at its disposal operationally. BrandWorks was tasked by the group to come up with a brand new Japanese concept, from strategy, brand design and interiors to conceptualising the menu and presentation. The brief was to create a holistic brand experience that engaged all five senses. The brief left the door wide open to push the boundaries and articulate Japanese dining in a thoughtful and provoking way that set it apart from others. With local consumers spoilt for choice, many brands have to give away food discounts and gimmicky marketing to attract their patronage. The challenge was to offer a destination dining concept that was focused on quality and experience, not on price.

Project details

▶ **Design:** Brandworks
▶ **Homepage:** https://www.brandworks.co
▶ **Area:** 230 m²
▶ **Location:** Jakarta, Indonesia
▶ **Photographs:** Sefval Mogalana, Kami Studio, Soemario
▶ **Branding and graphic design:** Brandworks

About us

BrandWorks are a multi-disciplinary team committed to developing strategic design solutions for FMCGs, Hospitality, Commercial Property, Digital Online brands and Fit-for-Purpose Retail Destinations. We have offices in Melbourne, Newcastle, Jakarta and Changsha, China. What makes us stand out from the crowd is that we've been there. From business management to ownership to operation, our experiences with and love for great design guides our process. We believe that our clients' success supports us and vice versa. So together, we're building a global business community curated for success with innovative and intelligent design. Advocating design-led thinking, we work to drive a positive impact on people, communities, and the economy so that we may all prosper in the future. We align with business leaders, entrepreneurs and changemakers who believe that their business can do better through design. By understanding their bandwidth, we set them up for growth and scale and meet at the intersection of transformation and success.

▶ Contacts e-mail: hello@brandworks.co / Call us: +61 3 9013 5333

From the insights of the market research, BrandWorks discovered that the majority of Japanese dining brands shared very similar design aesthetics and menus. This also extended to their brand identities and interior designs in delivering modern Japanese food. For SUSHI KAIYO, the opportunity was for the brand to take a different view towards Japanese culture, its food and origins.

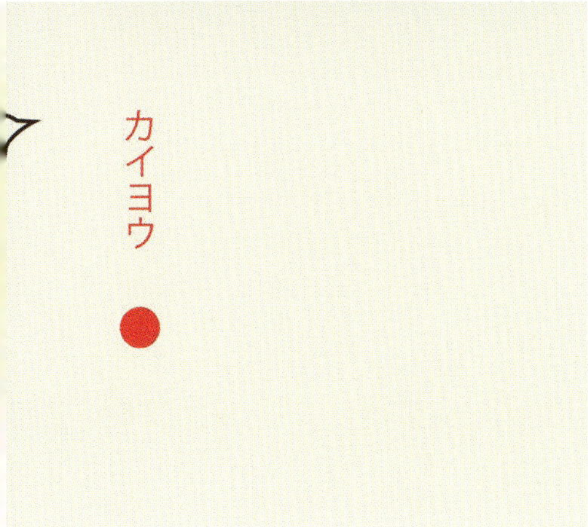

'Kaiyo' means ocean water. Inspired by nature in all its forms - Earth, Wind and Fire, KAIYO invites its audience to rediscover the forces of nature by showcasing how the food is sourced, prepared and served at the table. The brand and interiors elevate the mystery of the ocean with Japanese obsession with purity, perfection and craftsmanship. The design philosophy for the interiors takes on the Japandi (Japanese-Scandinavian) style, where the Japanese artisanship of making sushi is combined with contemporary Scandinavian modern finishing touches.

Separate from the FFE rationale, the joinery language is inspired by the Japandi design movement. All of the cladding detailing, joinery details and dowel work have been inspired by this movement that connects the minimalist design ethos of Japan called wabi-sabi, which loosely translates to understated elegance, and that of Scandanvia called hygge, which honours well-being and joy. Japandi movement was chosen to not only tap into the culinary offer but also works in with the forward movement of ensuring our spaces focus on wellness as a priority. We asked ourselves, how does the space make the person feel warm, comfortable and at ease.

The space was designed at the height of the pandemic and not knowing what the future held, design decisions were made in order to ensure our space was future proofed. Subtle elements like custom branded banners were mounted in between each table so if social distancing was paramount each patron can be separated by a purpose built japanese inspired curtain without compromising on the spatial aesthetics.

In early Feb 2020, the Founder and CEO of Boga Group briefed BrandWorks on their vision to establish a brand new Japanese food concept with a focus on quality sushi with an innovative twist. The engagement entailed developing a "high concept" experience with a cutting-edge brand, interiors and menu style for a proposed 300 sqm site within a shopping mall environment within Jakarta.

021

Aimed at the rising affluent middle class of young families, office workers and food seekers, the new concept was designed so that it could be taken to other cities and similarly populated regions in Indonesia. Significant resources and time was allocated on exploring the opportunities for the food, design aesthetic and the customer journey, which then extended to food consulting and menu design. This allowed BrandWorks to work closely with Boga Group and their operational team to explore, design and detail the entire brand and customer journey in a holistic way from start to finish.

024

We specialize in crafting mouth-watering meals and providing superb customer service. - A high concept Japanese bar and all-day dining.

Tero Brussels
Saint-Gilles, Belgium

A sustainable concept with seasonal, plant-based dishes – that is Tero, Esperanto for earth.

Tero Brussels: Rue St. Bernard 1, 1060 Saint-Gilles, Belgium / +32 2 347 79 46

A sustainable concept with seasonal, plant-based dishes – that is Tero, Esperanto for earth. After the huge success of Tero in Waver, it is high time to conquer all taste buds of Brussels and beyond. Tero entered a steaming affair with Creneau International. On the menu: the complete construction, project management and interior design, topped with a stunning centerpiece of a bar. Tero diners will have the seasons on their plate. The Tero farm, Ferme des Rabanisse, and collaborations with local businesses ensure ample supply of the freshest produce. This way, economical and ecological motives are united, as are health and gastronomy. First, we went to the restaurant in Waver to complete submerge ourselves into the Tero philosophy. The interior of the Waver location is beautiful yet simple. To translate it into a Tero Brussels we needed more – a tangible story. Everything revolves around the earth, Tero. Layered in all its aspects, from earthly elements and seasons to the surrounding cosmos. It translates into the Tero concept – from sow to harvest, from prep to serve. We wanted to implement these layers into the interior.

By placing an eye-catcher in the middle of the restaurant, we interconnect the different levels of the building. An impressive concrete bar was placed which immediately sets the tone. The bar perfectly illustrates the layering of the earth. "The concrete bar is a conversation starter, it calls for a response. This is exactly what I envisioned." - Maarten Groven, Senior Designer. The four seasons are entwined around a central column, the wine pillar. If you walk past the structure of slats, the image on them changes – a lenticular effect. From autumn to winter and spring to summer, to the rythm of the seasons. An optical effect from a Belgian landscape, a nod toward Ferme des Rabanisse. The ceiling was transformed into the cosmos. With the idea of connecting all spaces, we worked with concentric circles surrounding the central wine pillar. They provide the blueprint for all technical elements in the ceiling, including the custom-made luminaires.

Project details

▶**Design:** Creneau International
▶**Homepage:** https://www.creneau.com
▶**Area:** 273 m²
▶**Location:** Rue St.Bernard 1, 1060 Saint-Gilles, Belgium
▶**Photographs:** Creneau International

About us

We like to think before we do. That is why we start with the facts, then turn them around for you, offer you another take on things. Our job is to think harder. To see differently. And to treat you to fresh ideas.

You have plans—a hotel, a restaurant, offices, a whole city block, or maybe just your product's packaging. And you have questions—from how to master the master planning to which chairs, maybe even what should be on the menu. Before we do any designing or building, we strategise. We'll sit you down, dig for your reasons why, understand your world and come up with a blueprint for your business. The numbers, yes, but all the feels, too. Because a good strategy is one that rhymes ROI with speaking to your customer's heart.

▶Contacts e-mail: info@creneau.com / Call us: +32 (0) 11 24 79 20

RESTAURANT
TERO
COMPTOIR
ATELIER

winter
water
autumn
air
TERO
earth
spring
fire
summer

RESTAURANT
TERO
COMPTOIR
ATELIER

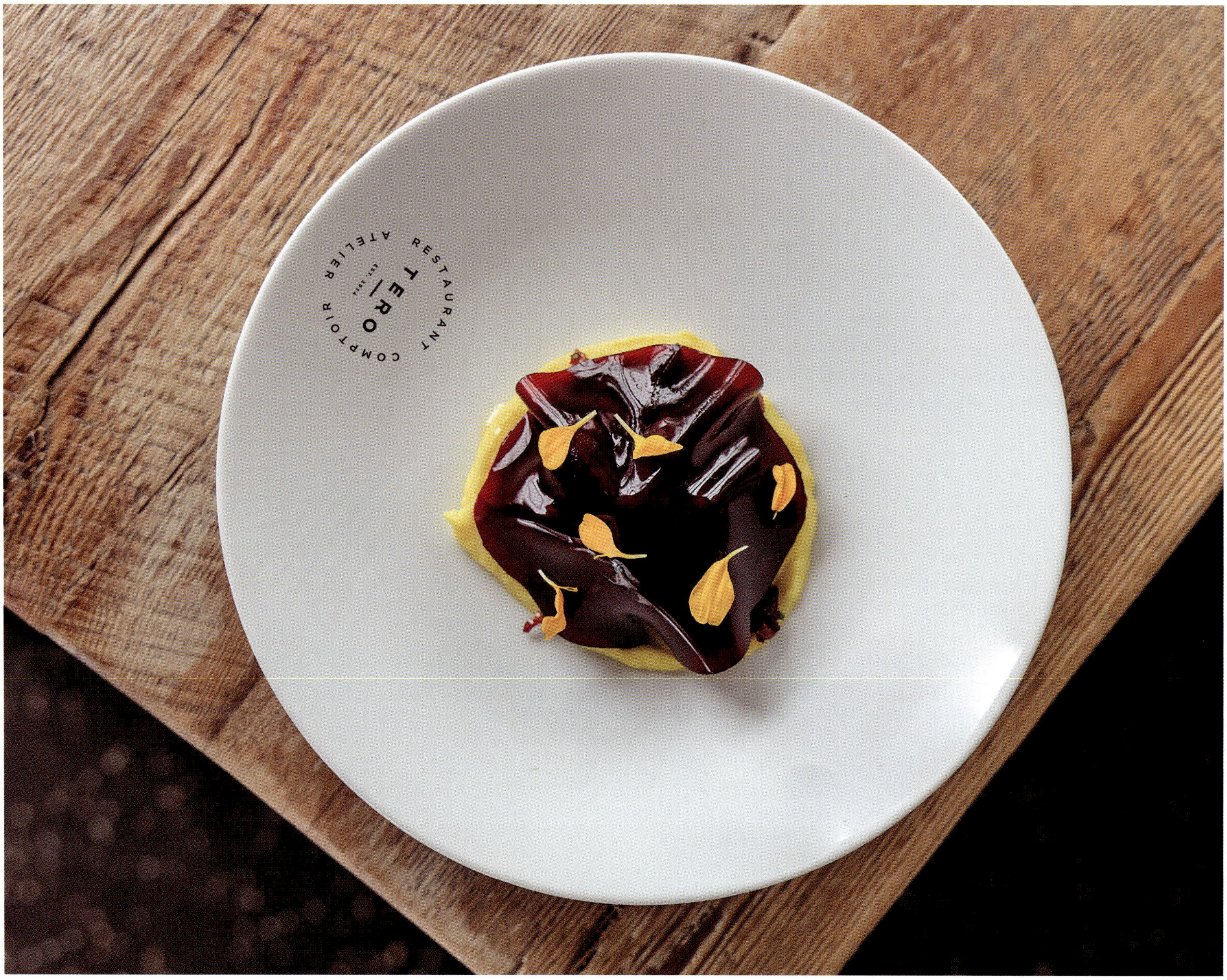

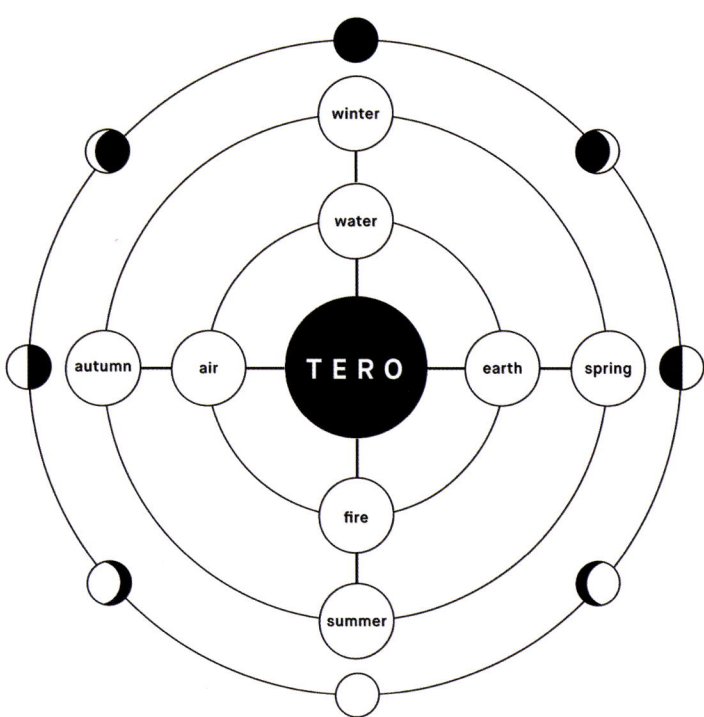

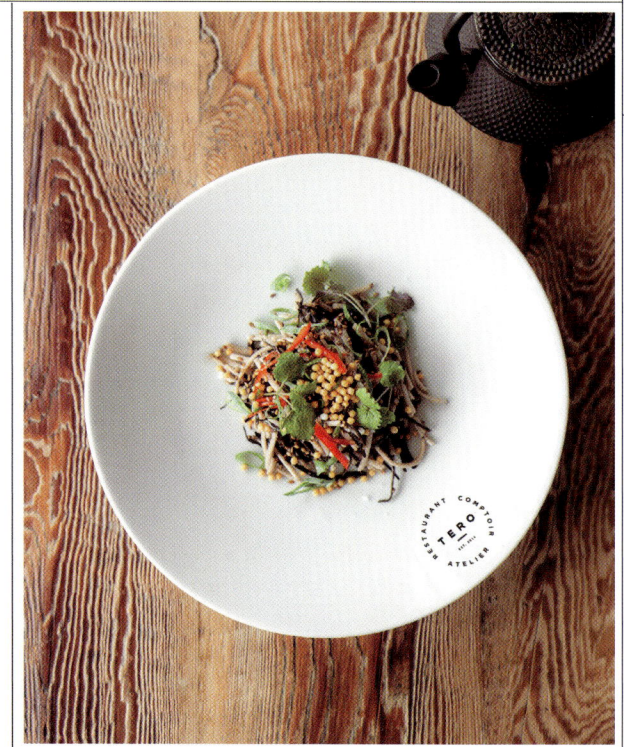

Tero entered a steaming affair with Creneau International. On the menu: the complete construction, project management and interior design, topped with a stunning centrepiece of a bar. Everything revolves around the earth, Tero. Layered in all its aspects, from earthly elements and seasons to the surrounding cosmos. It translates into the Tero concept – from sow to harvest, from prep to serve. We wanted to implement these layers into the interior. By placing an eye-catcher in the middle of the restaurant, we interconnect the different levels of the building. An impressive concrete bar was placed which immediately sets the tone. The bar perfectly illustrates the layering of the earth.

A central bar serves as the main focus with different layers and colours of concrete signifying the layers of the earth, and is based at two different levels. The lower section of the bar forms the foundation and is also the finishing of the lower cabinets. The terrazzo, quartz based floor provides a natural link to the earth while the ceiling is transformed into the cosmos, and roughly cut pieces of felt are layered on the walls to optimise acoustics.

035

Inspired by seasons and earthly elements, Creneau International designed and built the Tero restaurant in Brussels, with the four seasons entwined around a central column that houses the wine cellar. As guests walk past, the imagery on it changes to reflect the various seasons. A mood board with inspirational keywords also illustrates Tero's ideology. Based on the concept of connecting all the spaces, the design flows in concentric circles, surrounding the central pillar and providing a blueprint for all the technical elements in the ceiling, including the custom made luminaires.

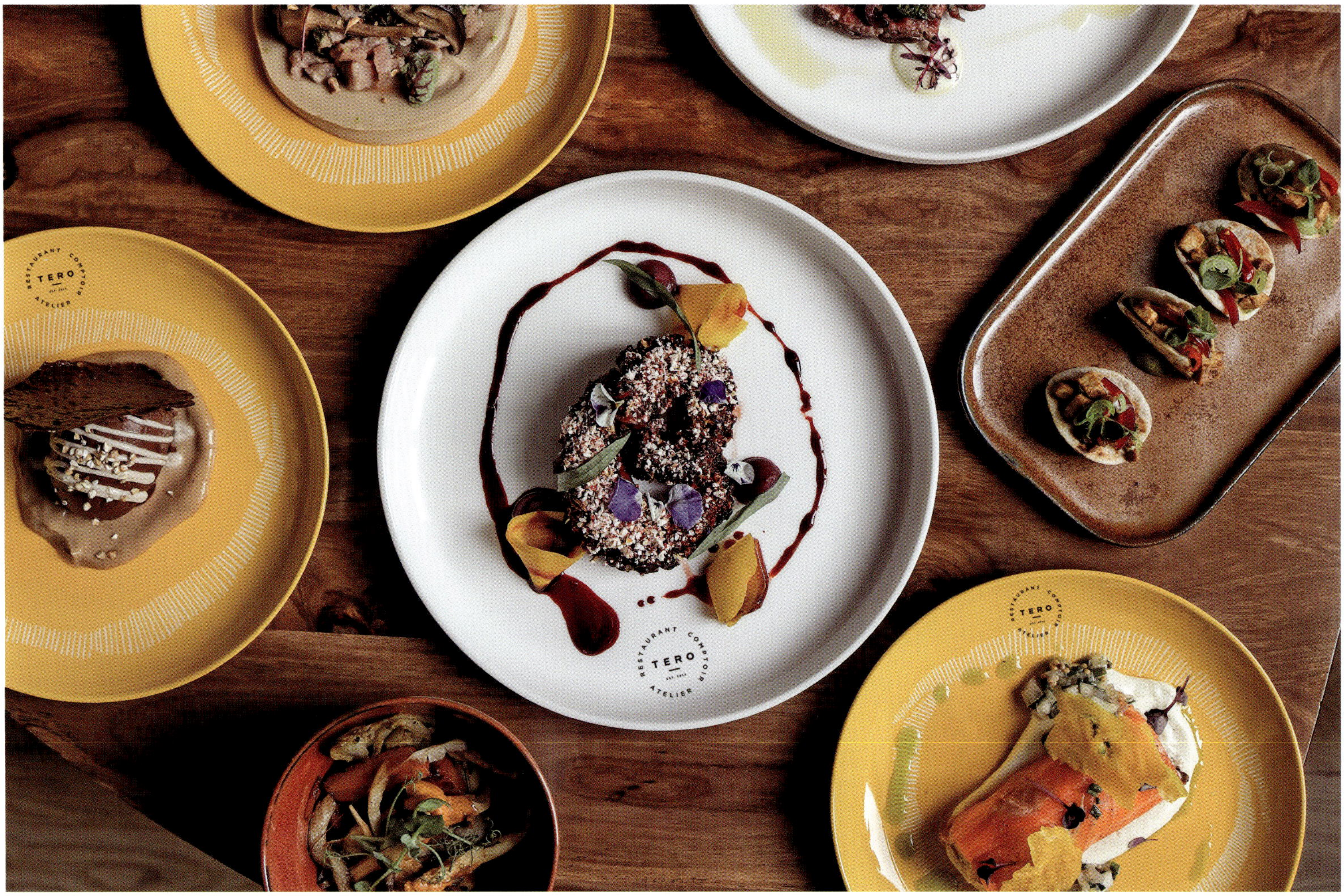
RESTAURANT
COMPTOIR
TERO
ATELIER

RESTAURANT
TERO
COMPTOIR

RESTAURANT
COMPTOIR
TERO
ATELIER

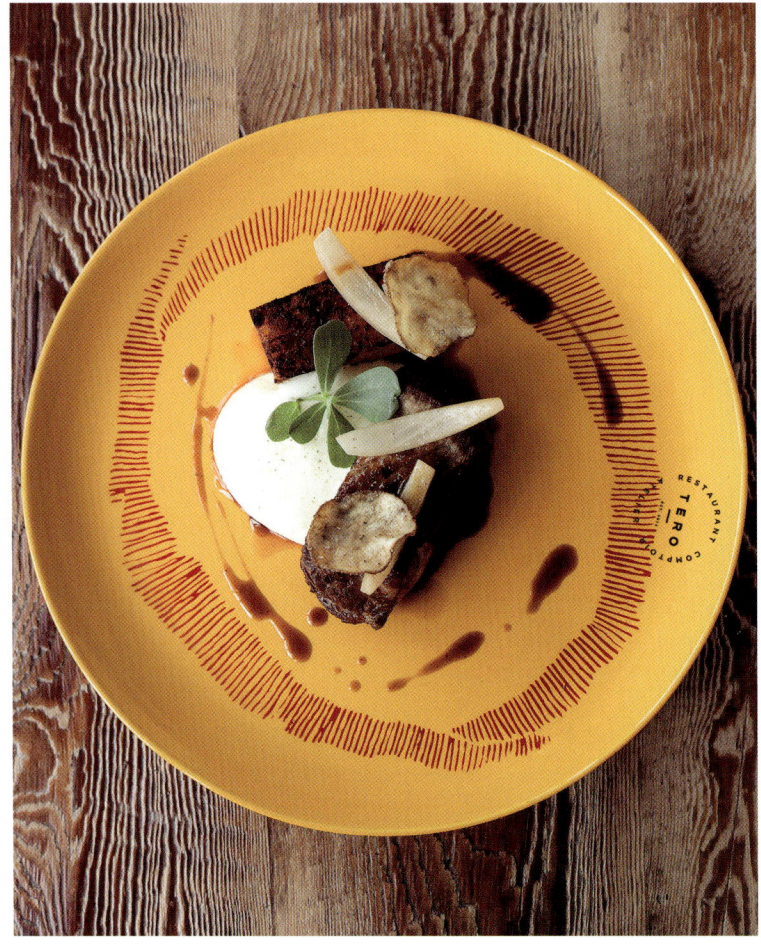

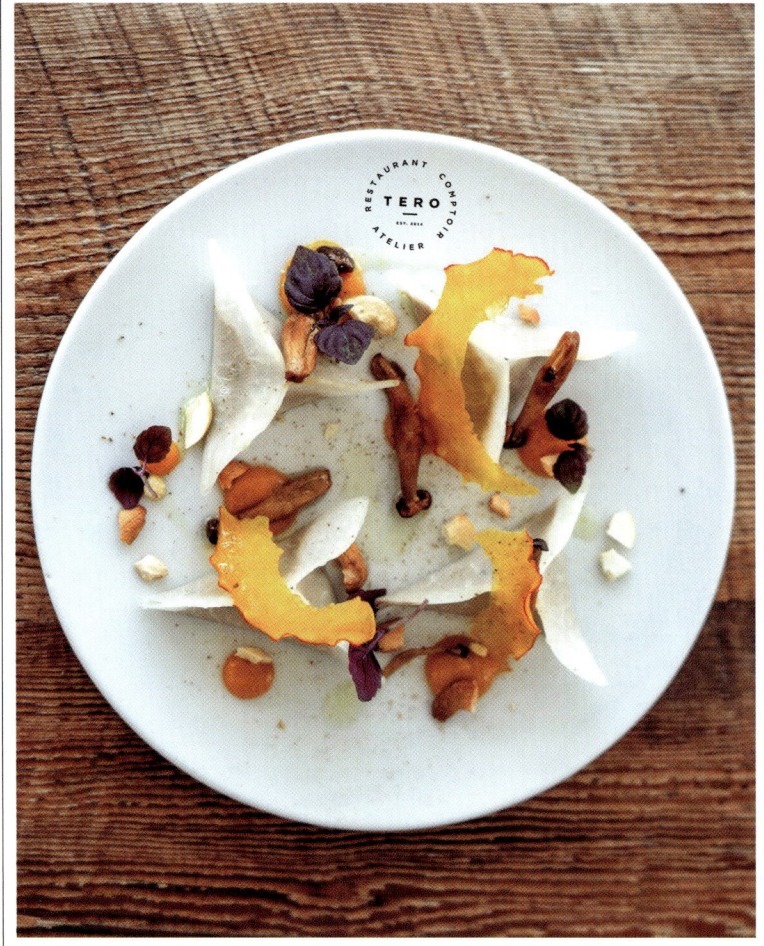

On the menu side, we ended up being quite disappointed with the veggies, which although they were perfectly cooked (the celeriac simply melted in our mouths), clearly lacked generosity and seasoning. On the fish and meat side, it was more of a success with the Lustin poultry with sage and pumpkin (13€), full of flavors and really tender, and the great home-smoked mackerel with olive sherbet (14€). The clams were delicious as well, but were also lacking that little something more in the sauce. We ended lunch with a vegan dark chocolate mousse (8€), with a texture as dense as the one of a ganache, without neither eggs nor milk, but with the intense flavor of dark chocolate. A very impressive dessert and know-how which clearly made up for the rest of the meal. A place to come back to in a while, when the team in the kitchen is a little bit more together and ready to match the first venue's reputation.

By placing an eye-catcher in the middle of the restaurant, we interconnect the different levels of the building. An impressive concrete bar was placed which immediately sets the tone. The bar perfectly illustrates the layering of the earth. "The concrete bar is a conversation starter, it calls for a response. This is exactly what I envisioned." - Maarten Groven, Senior Designer.

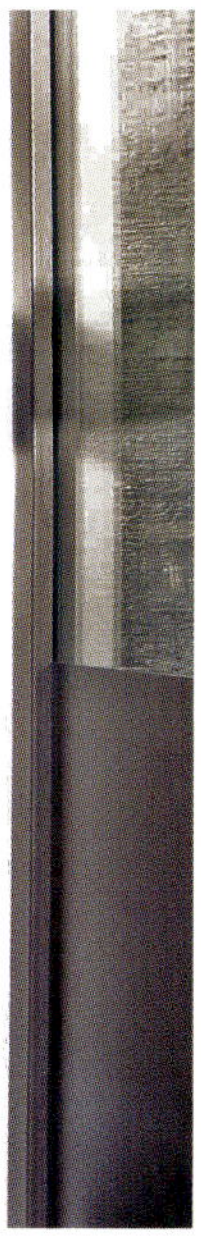

▶Sushi & Co. : Helsinki, Finland / +32 2 347 79 46

Sushi & Co.

Helsinki, Finland

Sushi & Co. is a restaurant and cafe on-board a cruise ship taking guests to destinations along the Baltic Sea.

Helsinki-based design agency Bond created this clever new visual identity for Sushi & Co, a sushi restaurant on a Baltic cruise ship. "Bond designed a simple and clever logo and a brand identity. The new design incorporates Scandinavian elements with a sophisticated color scheme and oceanic symbols." Sushi & Co. is a restaurant and cafe on-board a cruise ship taking guests to destinations along the Baltic Sea. It has a modern interior design that mixes dark and light wood furniture, features warm low hanging lights, organic patterned upholstery, cool grey walls, exposed brick panels, slate floors and a visual identity developed by Helsinki based graphic design studio Bond. Extending across menus, napkins, uniforms and signage, Bond's visual identity leverages association, familiarity and the unexpected, through colour, form and type, to introduce a current Scandinavian simplicity to a more detailed but equally precise interior.

Where the wave and scale-like pattern, salmon red colour and fish iconography embraces familiarity and association within the context of a seafood restaurant, all the more relevant because of the restaurant's cruise ship location, the monolinear, monospaced and condensed qualities of the logotype, punctuated by a moment of play within the ampersand, is unusual and unexpected.

Although mechanical and impersonal across the menus, type is well integrated into environment through wood and illuminated signage. Much like the fish within the ampersand, the use of a wood grain across the surface of monospaced type is distinctive in its contrast of ornamental flourish and typographical utility. This contrast continues through to a white paper and black ink economy alongside panels of a Pantone red that interrupts a dark interior with a moment of colour, in the same way the geometric wave pattern breaks from organic material texture.

Project details

▶**Design:** Bond
▶**Homepage:** https://www.bond-agency.com
▶**Area:** 169 m²
▶**Location:** Helsinki, Finland
▶**Photographs:** Angel Gil

About us

BOND is independent brand and experience design agency founded in Helsinki 2009. We are a versatile and collaborative team of designers, technologists, strategists and storytellers. Five studios. 55 people. 19 nationalities. We marshal expertise across disciplines to craft the fusion of experiences your customer has with you. Simple Wins is our business, brand and design philosophy. It is what guides what we do for our clients and how we act as an agency. Keep it simple. Know who matters and what they want. Then design brands and experiences that perfectly serve them, everywhere, always. Be creative and bold. Don't deviate or be diverted. In our complex world, this is how we win.

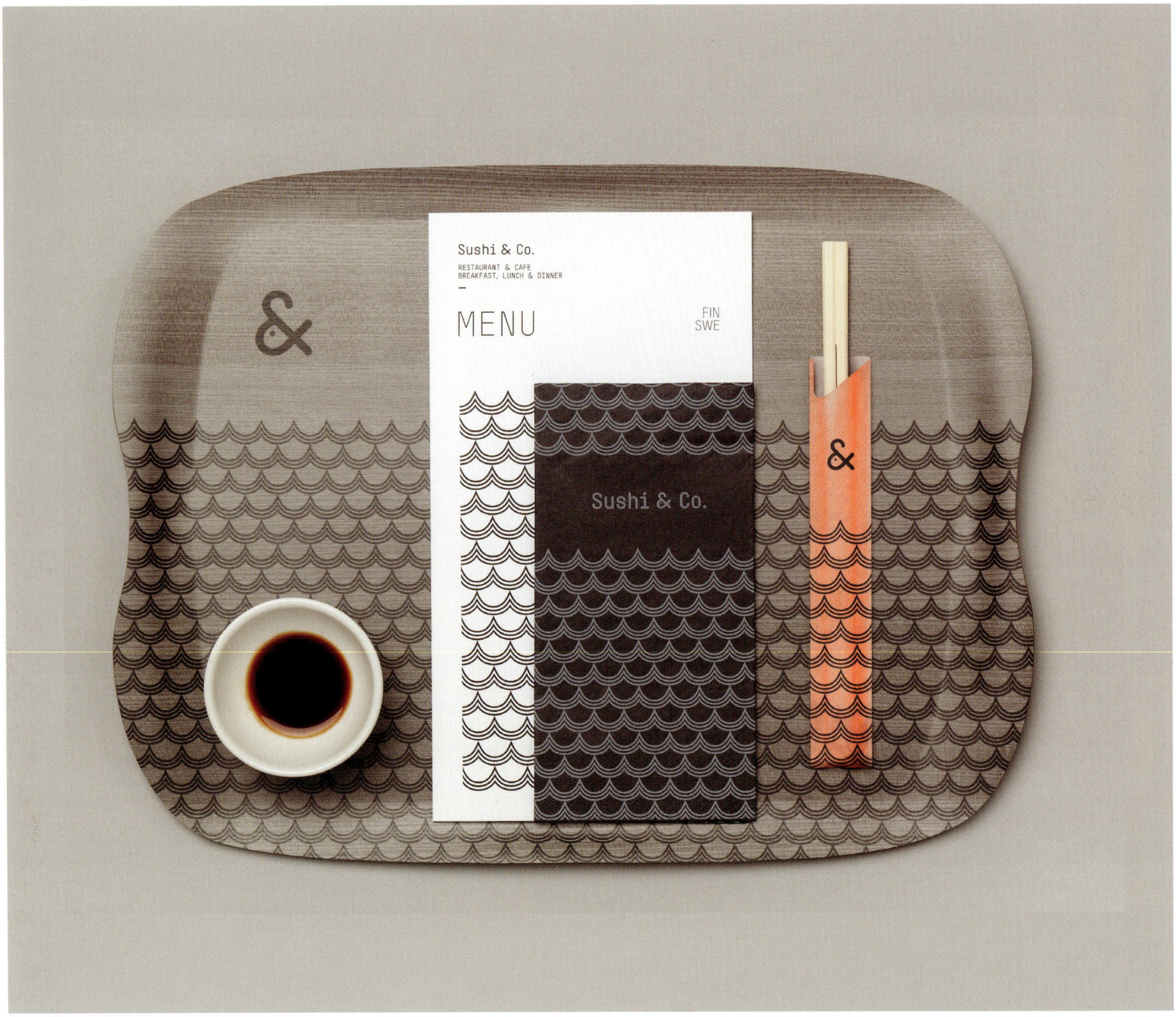

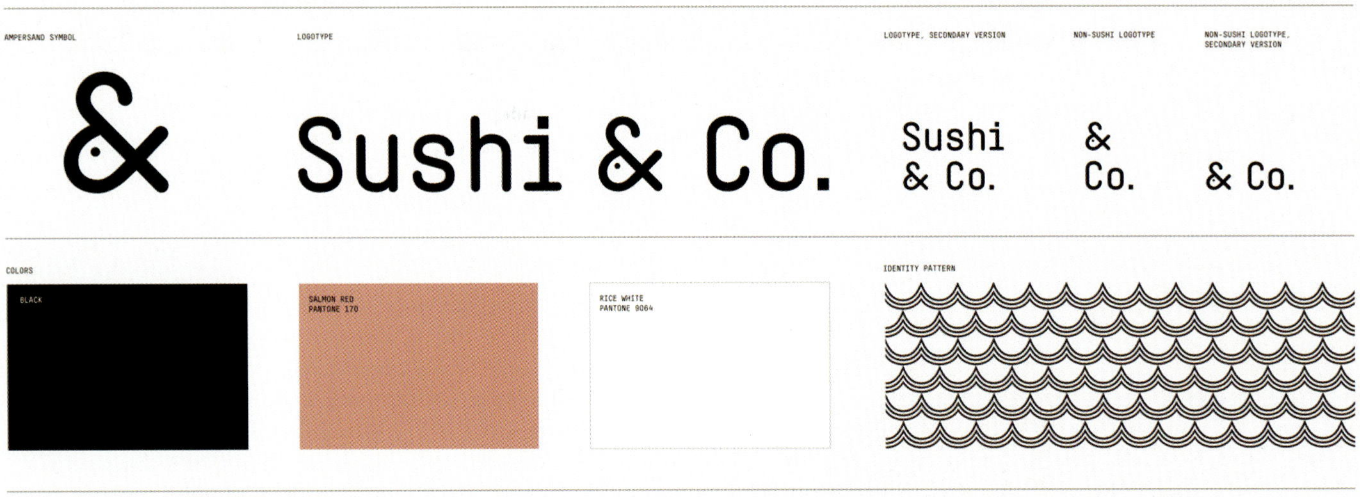

AMPERSAND SYMBOL	LOGOTYPE	LOGOTYPE, SECONDARY VERSION	NON-SUSHI LOGOTYPE	NON-SUSHI LOGOTYPE, SECONDARY VERSION
&	Sushi & Co.	Sushi & Co.	& Co.	& Co.

COLORS

BLACK	SALMON RED PANTONE 170	RICE WHITE PANTONE 9064

IDENTITY PATTERN

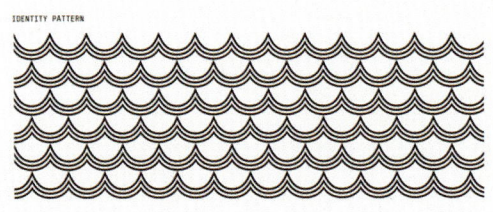

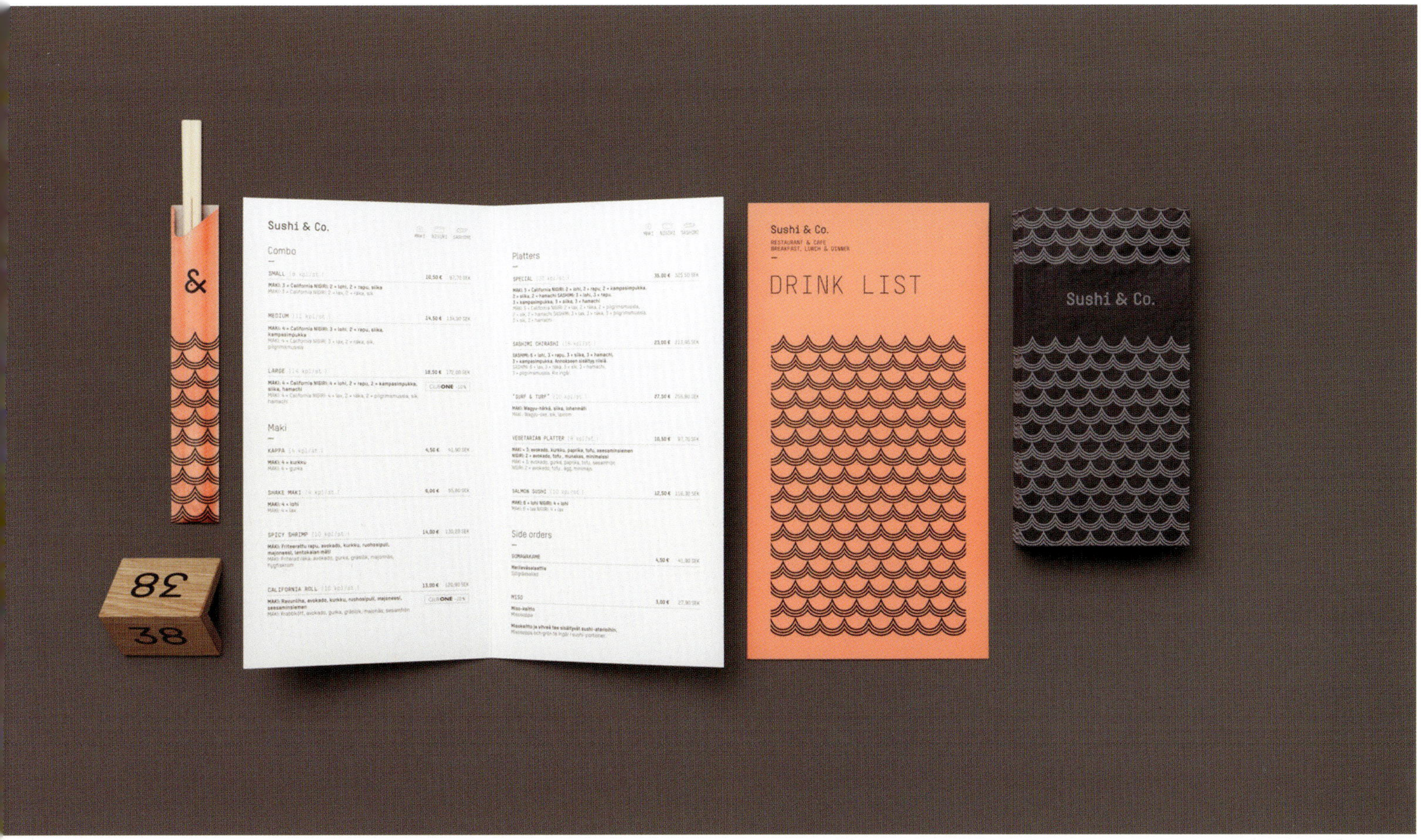
Sushi & Co.
Combo
Maki
Platters
Side orders
Sushi & Co.
DRINK LIST
Sushi & Co.

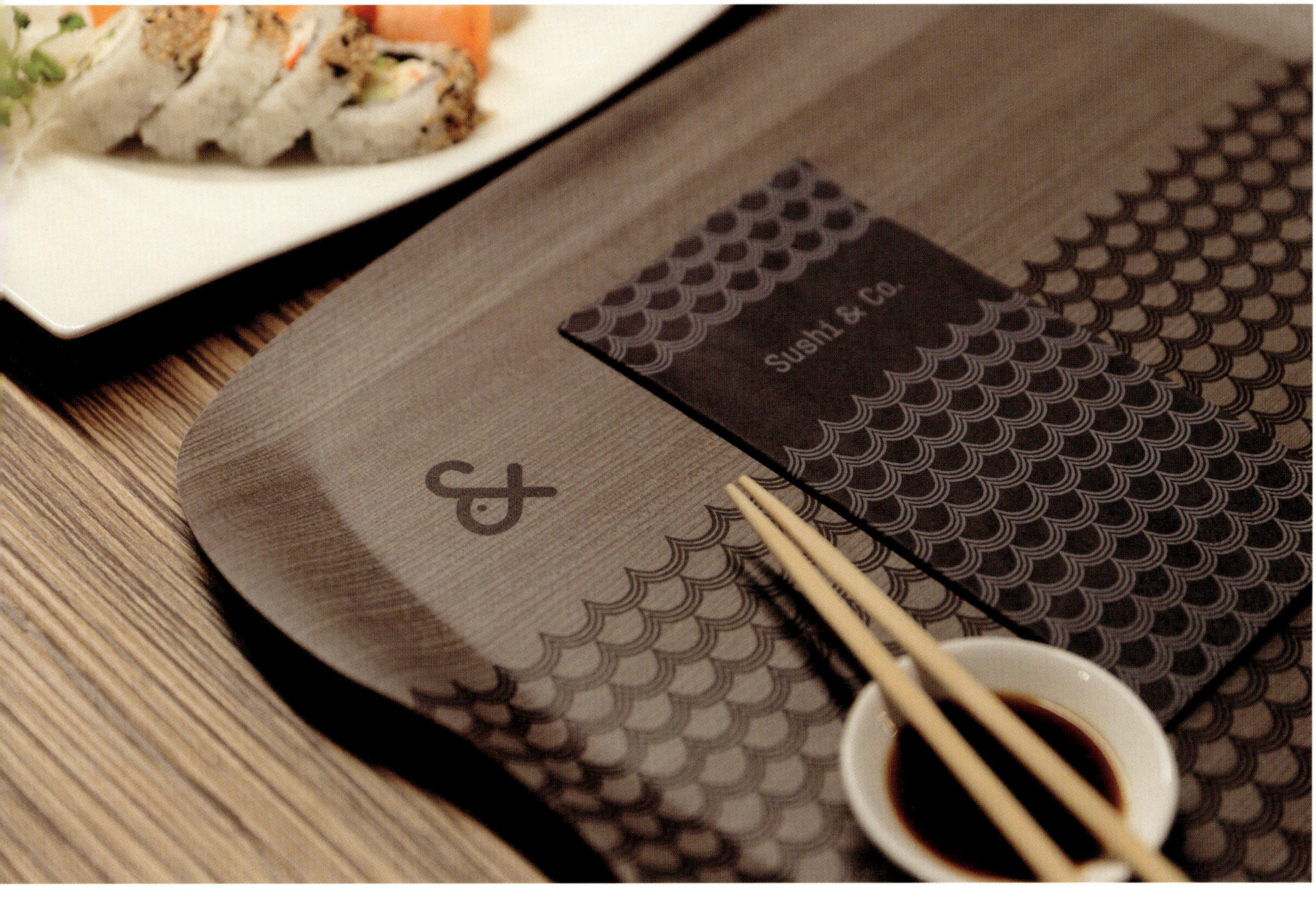
Sushi & Co.

& Sushi & Co.

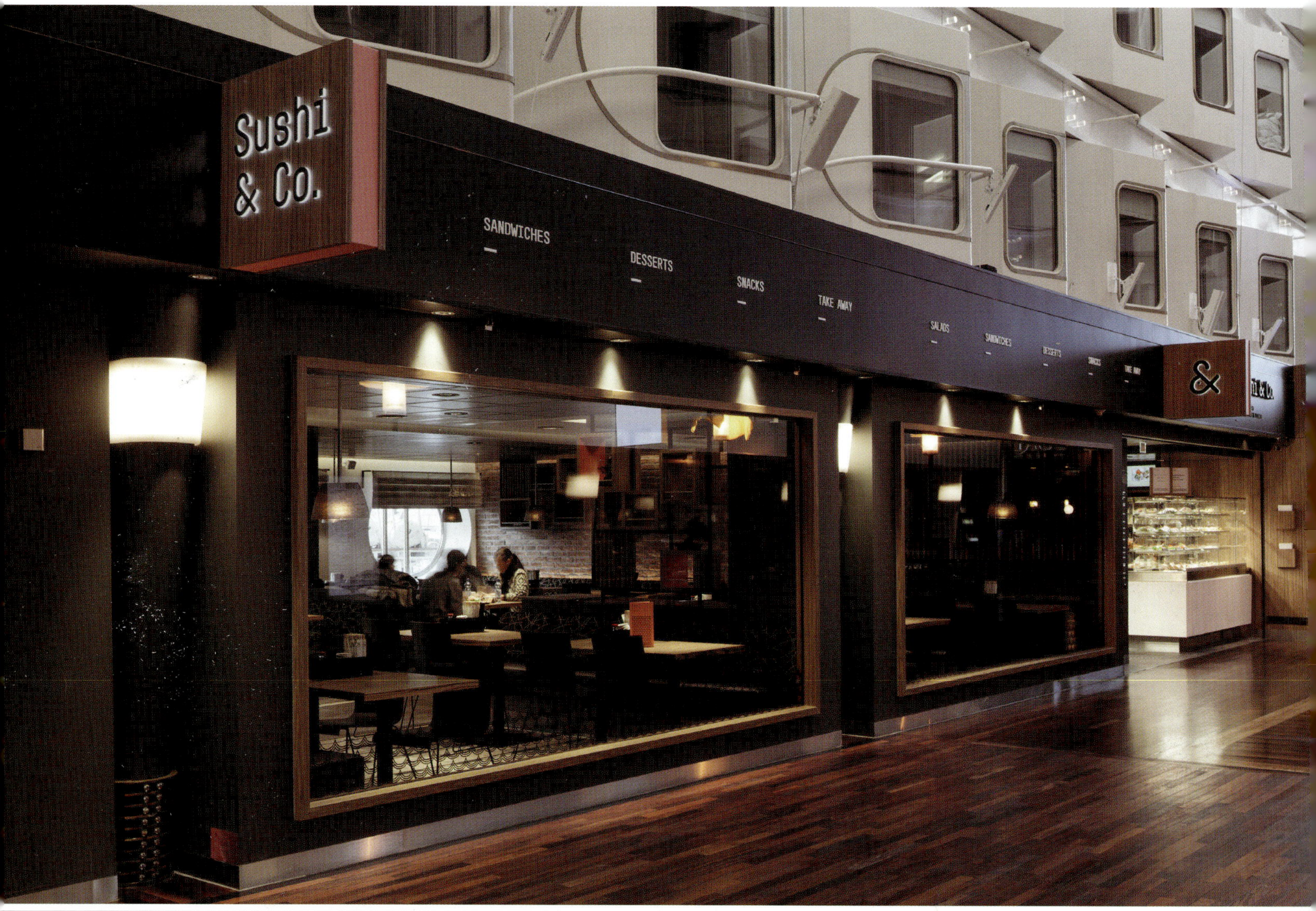
Sushi
& Co.
SANDWICHES
DESSERTS
SNACKS
TAKE AWAY
&

&
SALADS
SANDWICHES
DESSERTS
SNACKS
TAKE AWAY

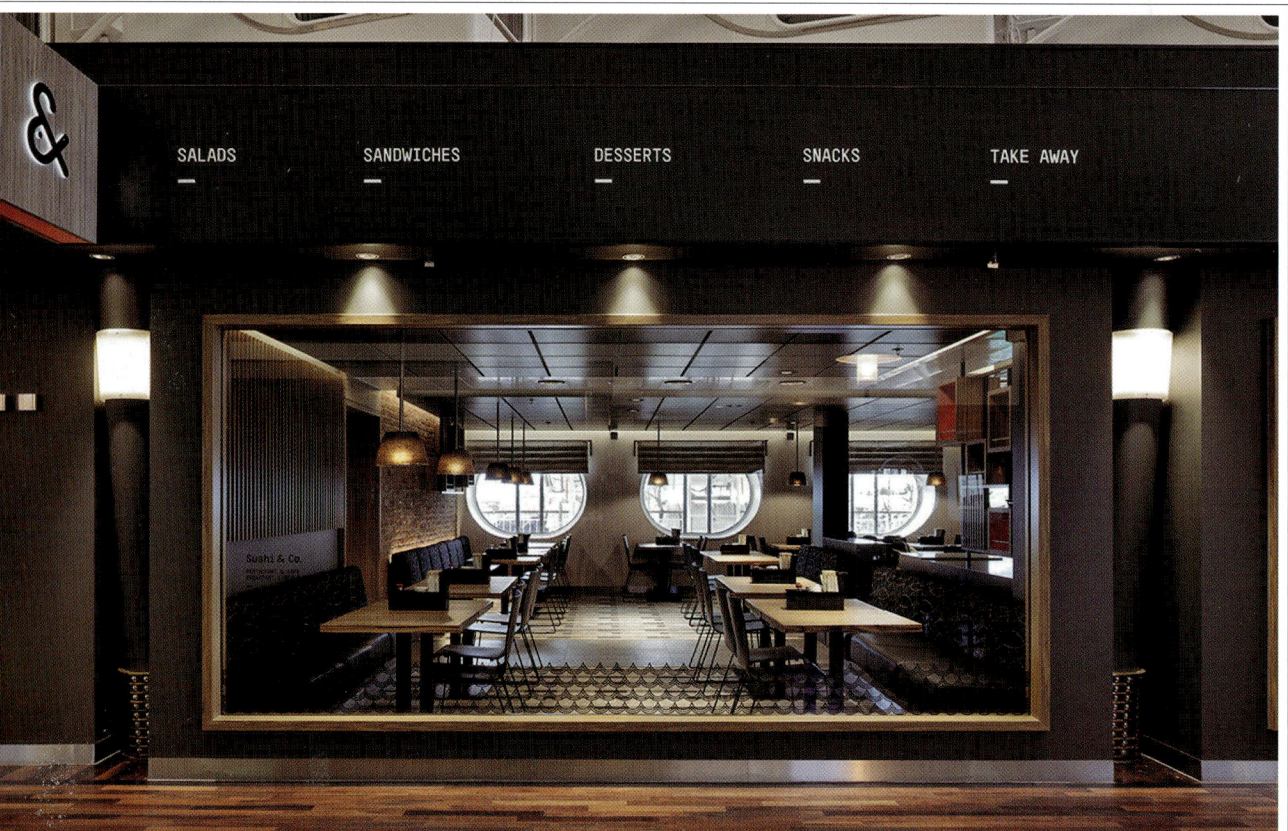

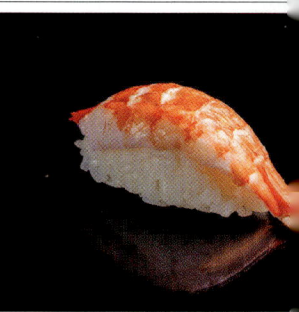

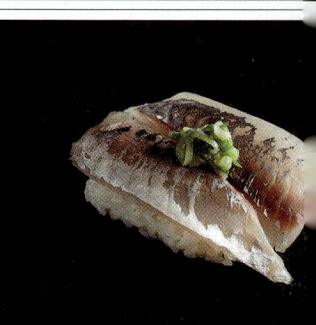

Kyiv, Ukraine
Porto Maltese

The restaurant is located in one of the central districts of Kyiv, among busy traffic routes.

The main idea is to create the atmosphere of a Mediterranean town. To take the visitor away from the city's hectic life for a leisurely visit, immersing himself in the atmosphere of a small port, with the opportunity to dive into the depths of the sea. Here you can warm up in the warm company of close friends, share a delicious dinner, soak up the blueness of the waves and emerge refreshed in anticipation of the next visit to the port. The complex shape of the restaurant premises immediately suggested the motive of the future design – two halls, like the bays of a port city, flow into each other. They are united by a fish market with fresh catch just under the sail of fishermen and completed by a summer terrace – an exit from the port to the city. And there is nothing better in the traditions of the Mediterranean than combining fresh fish with a family winery.

The interior is soaked in deep sea blue with bright snow-white accents: sails and tablecloths. In addition, many details create the atmosphere of the port. Jellyfish lamps slowly float above the bar against the background of the rusty ship, waiting for guests. A wall with an image of the terrible lord of the seas, the whale from Konrad von Gesner's "Animal History". It was tamed to protect the exquisite treasures of our restaurant – a collection of bottles of selected wine, which, according to legend, is created in the nearby hills.

On the ceiling connecting both halls, sea waves sway in smooth curves, playing with rays of warmth, like sunlight from many light bulbs, twist into a vortex, drawing the visitor to explore the entire space, explore both halls. Rope columns – carefully stacked gear, without which the life of the port is impossible, attract attention with the skill of their execution.

Project details

▶ **Design:** loft buro

▶ **Homepage:** https://loftburo.com

▶ **Area:** 238 m²

▶ **Location:** Kyiv, Ukraine

▶ **Photographs:** Serhii Polyushko

About us

loft buro, est. 2001, is a creative team, which consists of professional architects, designers and painters. During this period of time, many projects in interior design and architecture were made by the team. Our main task is the creation of a harmonious, comfortable and cosy space that displays the inner world of a person and creates the perfect mood for all to enter it. Our main task is the creation of a harmonious, comfortable and cozy space that displays the inner world of a person and creates the perfect mood for all enters it

A communal table resting on a coral reef is for a large team that came to town to celebrate a successful trip loudly. Behind them is a wine cabinet with a shimmering surface of mesh facades. We have created a cosy corner for a family that wants to be alone in the company of goldfish, make a wish and get pleasure from the evening – visual and taste. Interior details are not immediately revealed even to a curious visitor, and the immeasurable depth of flavours calls to return again and again.

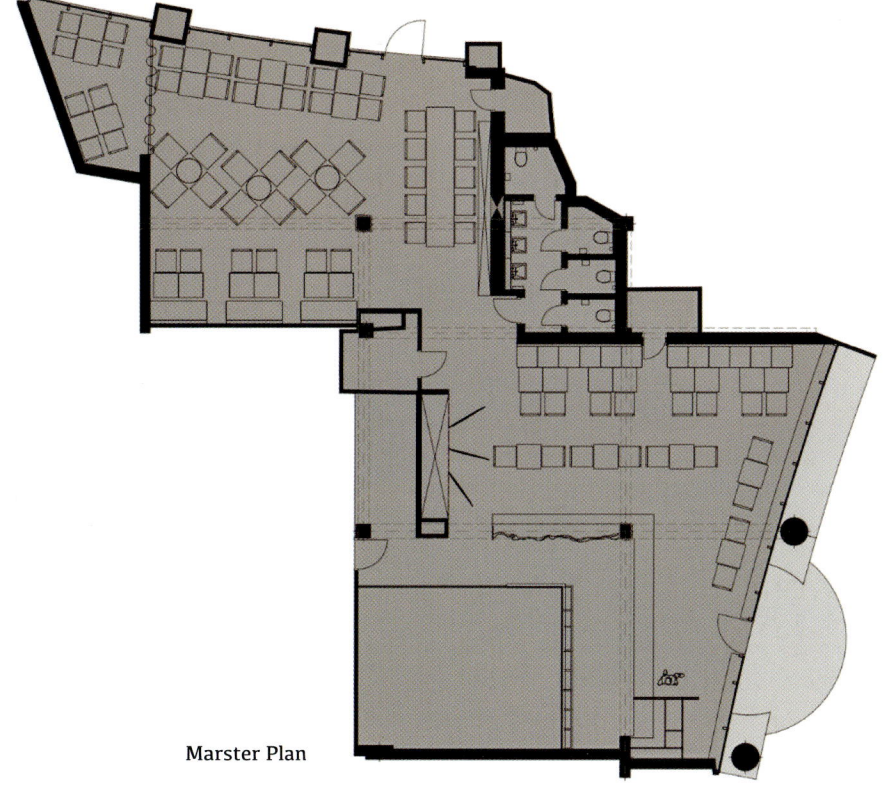

Marster Plan

Here, you can relax with friends, enjoy a delicious dinner, and soak up the blueness of the waves. The interior is beautiful and inviting, with deep sea blue and snow-white accents, jellyfish lamps, a wall with a whale mural, sea waves on the ceiling, rope columns, a communal table, and a wine cabinet. Every detail has been carefully crafted to create a unique experience, so come explore Porto Maltese and get lost in its immeasurable depth of flavors!

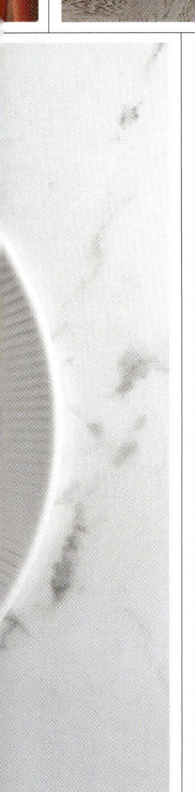

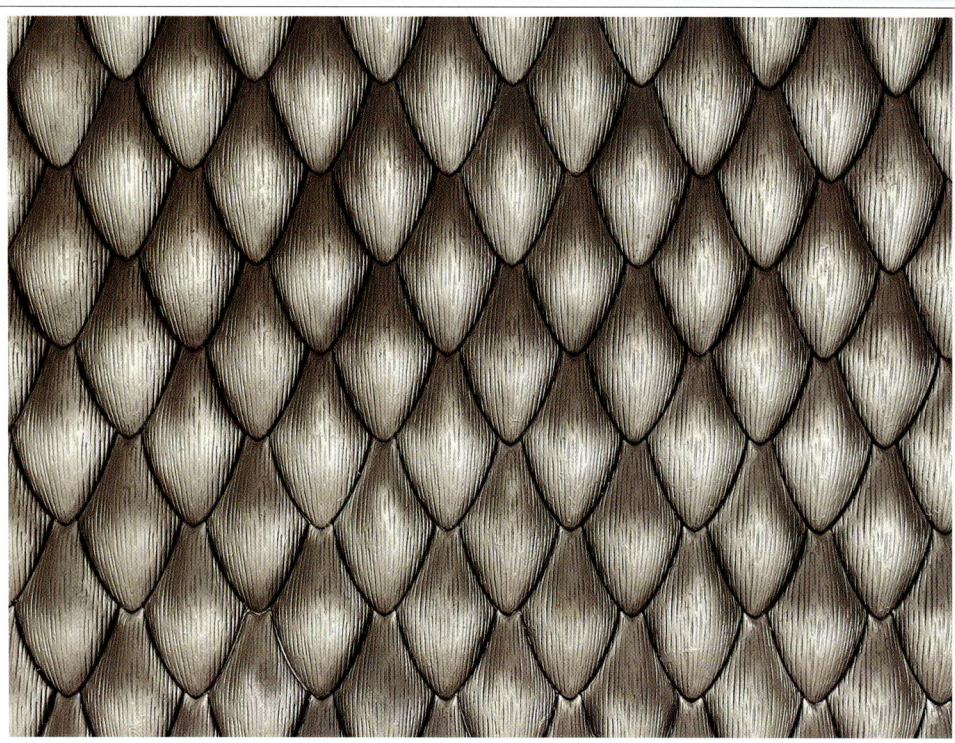

The interior details of this Mediterranean-style restaurant are sure to create an unforgettable experience. With its combination of fresh seafood, special wines, and unique design, this restaurant is sure to leave guests in awe and eager to return.

▶Osteria Betulla: Saint - Petersburg, Russia / +7 921 181-94-60

Saint - Petersburg, Russia
Osteria Betulla

The concept of osteria Betulla relies on simple Italian food in the original version with an emphasis on high quality of products.

Osteria 'Betulla' is the second project of a talented chéf Arslan Berdiev, the founder of the sensational restaurant Birch. Both 'Birch' and 'Betulla' are translated into Russian as "a birch". But whereas 'Birch' is based on pan-European culinary traditions, 'Betulla' focuses on the Italian cuisine. The concept of osteria Betulla relies on simple Italian food in the original version with an emphasis on high quality of products. An important feature of all Arslan's projects is perfectionism and unexpected serving of dishes, in a manner surprising for the guests, which is reflected in the osteria' s cuisine. From the very beginning, we were orientated at the image of an "Italian dining hall", but in a very minimalistic and pure manner.

We knew how sacred the restaurant team felt about their work, so we wanted to translate that feeling into a visual image. This is how we came to the idea of a certain "food temple" – a place resembling an uncluttered, minimalistic European chapel flooded with light. At the same time, we wanted not just to recreate a classical Italian interior, but to bring the modern spirit of Italy to St. Petersburg, using traditional colours, shapes and materials.

In working over this project, we used the architectural dramaturgy characteristic of traditional churches. The guests get into the space through a tapering ensconced, shaded entrance. They enter into the first, dimly-lit small hall, and further – into the second hall flooded with light. This way the guests proceed smoothly from shadow to light. We widened all window openings in the hall (which is located in the basement, below the ground level) and lowered them to the seating level, for the hall to be better exposed to light and not be perceived as an underground vault.

Project details

- **Design:** DA bureau
- **Homepage:** https://da-bureau.com
- **Area:** 86 m²
- **Location:** Saint - Petersburg, Russia
- **Photographs:** Sergey Melnikov

About us

The world driven by the conservative professionals has existed for way too long. Our hierarchyless approach is what forms us. Listening to the passionate and fresh-thinking talents is a crucial part of our culture. And this significantly influences our output and, therefore, the environment and society surrounding us.

we believe that innovative thinking, honesty and passion is a mind-shifting combo that brings good to everything we produce. While actually implementing everything we do is crucial for us.

We placed the elements referring to Catholic aesthetics practically everywhere. The centrepiece of the first hall is the metaphorical altar – a large table where the pasta chéf works. All the tables are turned towards him, which creates the effect of theatrical performance and allows the guests to watch the pasta making mystery. The zest of the second hall is the church pews in the centre. At the same time, the accents are made on the olive tree and the three kiots (a kiot is a niche traditionally housing icons of saints). Since the Italian cuisine has its own holy trinity – wine, olive oil and thyme – we placed them in the niches.

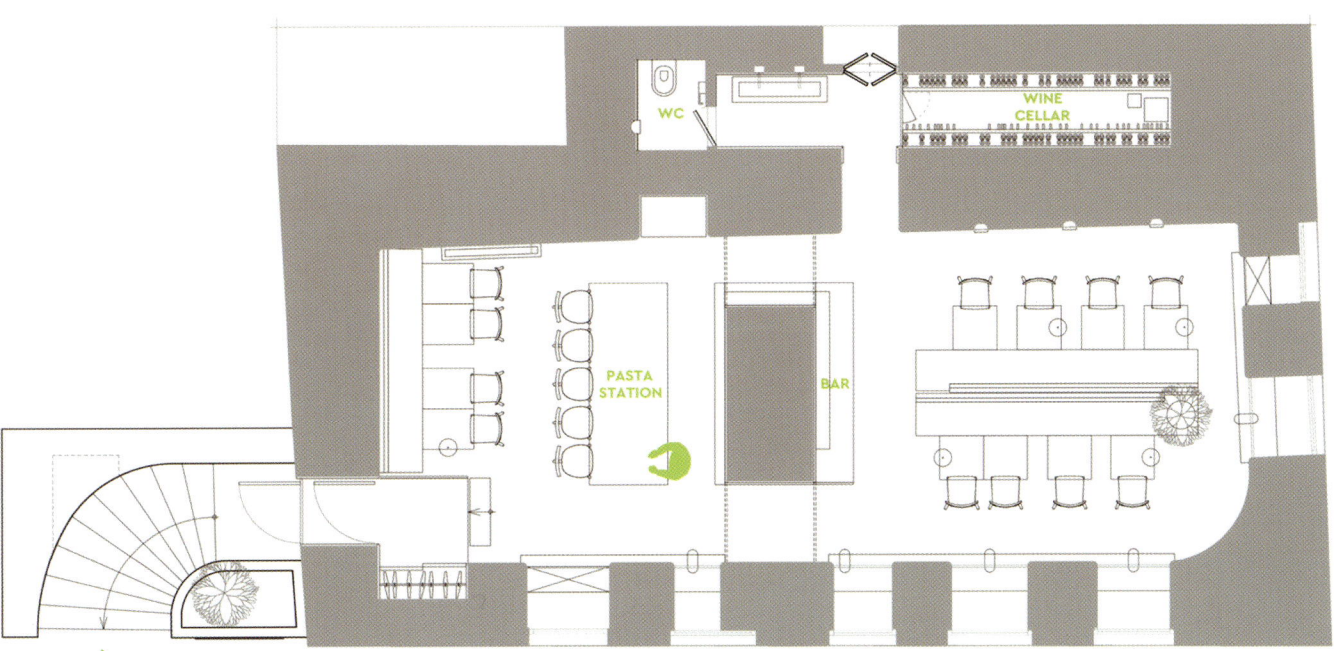

Floor Plan

083

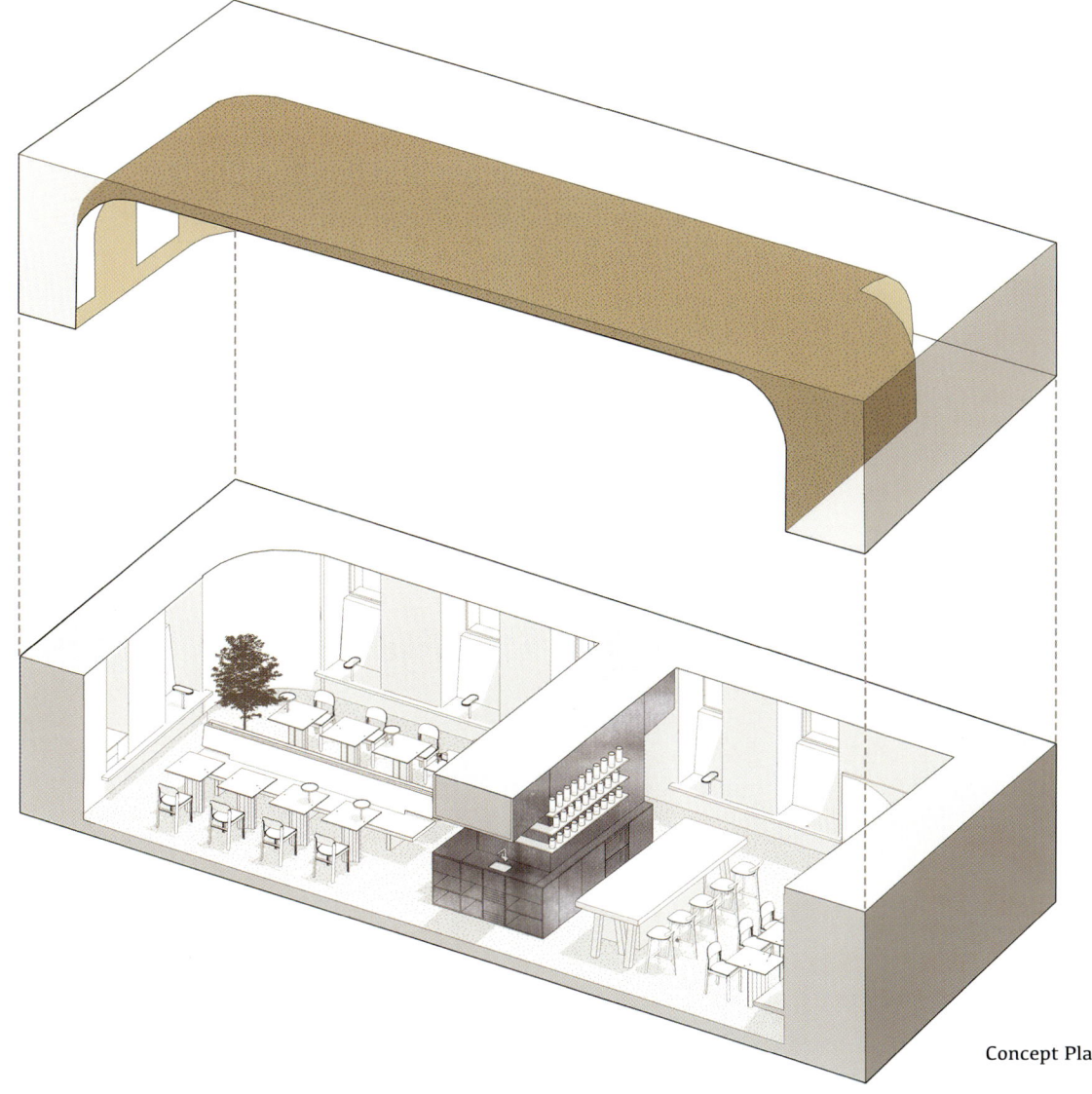

Concept Plan

The drinking fountains, common in Italian cities, took the form of a wine cooler in the first hall and a sink in the rest room area. Since Italy is famous for its wines, we couldn't do without the wine cellar which is effectually positioned in the under-staircase space. Owing to the light colour play and the mirrors on the walls, we were able to visually flatten the geometry of the cellar. We placed the elements referring to Catholic aesthetics practically everywhere. The centrepiece of the first hall is the metaphorical altar – a large table where the pasta chéf works. All the tables are turned towards him, which creates the effect of theatrical performance and allows the guests to watch the pasta making mystery.

The zest of the second hall is the church pews in the centre. At the same time, the accents are made on the olive tree and the three kiots (a kiot is a niche traditionally housing icons of saints). Since the Italian cuisine has its own holy trinity – wine, olive oil and thyme – we placed them in the niches. The drinking fountains, common in Italian cities, took the form of a wine cooler in the first hall and a sink in the rest room area. Since Italy is famous for its wines, we couldn't do without the wine cellar which is effectually positioned in the under-staircase space. Owing to the light colour play and the mirrors on the walls, we were able to visually flatten the geometry of the cellar. In addition to introducing the Italian flair, one of the main objectives was to transform the existing dark space of the basement into a well-lit airy space that would not oppress the visitor. The task seemed very difficult, even impossible at some point. The main architectural device in working with the interior space was the vaulted ceiling.

▶ A Japonesque Chef Restaurant: Tel Aviv, Israel / +7 921 181-94-60

Japonesque Chef Restaurant
Tel Aviv, Israel

A is a Japonesque Chef Restaurant located on the second floor of Azrieli-Sarona Tower in Tel Aviv.

The restaurant's chef, Yuval Ben-Neriah, envisioned it as a culinary institute. It was destined to offer a place to experience rich, timeless knowledge well blended with the chef's personal interpretation and daring culinary innovations. His idea to fuse Japanese and Western cultures into a slick lively elegance drove us to draw upon classic architectural idioms of form and matter in these two cultures, and re-sculpt them to create a calm yet assertive space and experience.

The entrance is constructed with the idea in mind of Japanese ceremonies and the experience of discovery. As the main entrance to the restaurant is through the tower's mall, we wished to create an opportunity that allows the guests to leave the hustle & bustle of the crowds behind and focus upon the new experience that lies beyond the door. A curving wall greets A's guests and gently directs them to the door located at the end of an elongated vestibule. Upon the passing of the threshold the story of A begins to unfold. The curving wall is joined by a straight perpendicular wall and together the two lead the guests inwards to be met by the host. At this point the curving wall is detached from the ground and turns to encircle the private seating space, while the visitor is directed to the main dining space, where the view opens and the totality of the architectural gesture is revealed in its fullest.

Project details

▶ **Design:** Baranowitz & Goldberg Architects, Pitsou Kedem Architects
▶ **Homepage:** https://www.baranowitz-goldberg.com
▶ **Area:** 550 m²
▶ **Location:** Tel Aviv, Israel
▶ **Photographs:** Amit Geron

About us

Having opened in 2017, Baranowitz & Goldberg partners Sigal and Irene have been collaborating as colleagues in many different commercial design projects. After a few years of successful joint work, the two architects established their own practice, paving a new way based around their diverse expertise. Irene & Sigal's comprehensive experience in different fields of design allow the studio's work to vary and include all scales. Whether it be a product or a building, creating a story that will ignite the creative process is the leading principle of their work.

The studio's projects are characterized with a playful balance between art and design while adhering to strict yet poetic principles of proportion, and a true integration between conceptual thought and functionalism. When created within a strong conceptual framework, creativity can turn to unpredictable yet coherent directions. The studio's goal is to allow each project to pave its way toward a unique architectural interpretation.

The straight wall becomes the main architectural feature and assumes the role of defining and sculpting the main dining hall. It is the partition separating the kitchen from the restaurant and it stretches all along the hall. It is the base of the halved-arch that governs the entire space with one single brush stroke. This classic architectural form was drawn directly from the existing proportions of the given space. Being long, narrow and tall, it evoked the associations of classic colonnades and churches. The fact that a long narrow terrace is stretched all along it, offering a beautiful view of Sarona-Park, made it imperative to direct the view in that direction and establish a strong connection between the indoor and outdoor.

This main partition that defined the dining hall is horizontally divided in three, where each part fulfils a different role: A graphic composition in the language of the half-arch realized in turquoise copper patina cladding embellishes the base of the partition. The patina recalls classic domed edifices and sculptures adorning western cities and elegantly expressing the passage of time. Above this base stretches a long narrow window exposing the industrious activity and the creative culinary force beyond the wall. Above this middle layer the halved-arch juts out to define the experience from above.

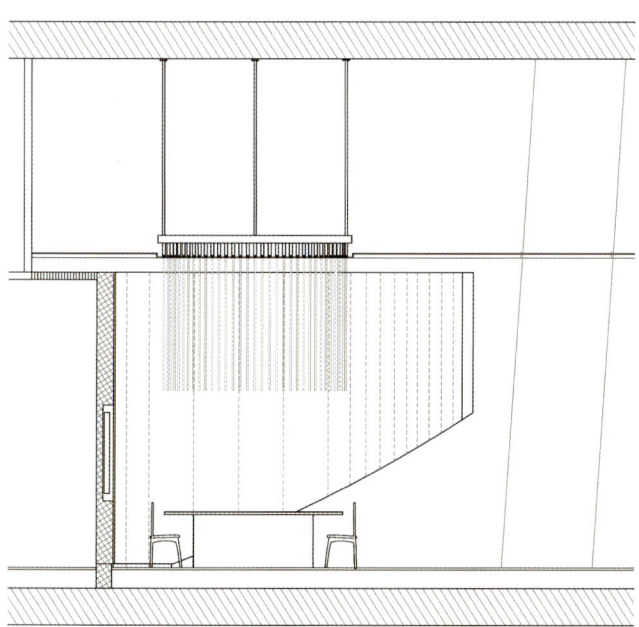
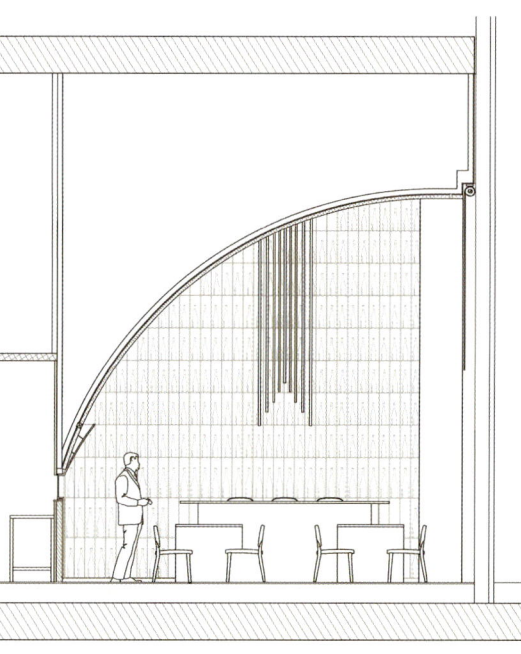

Section

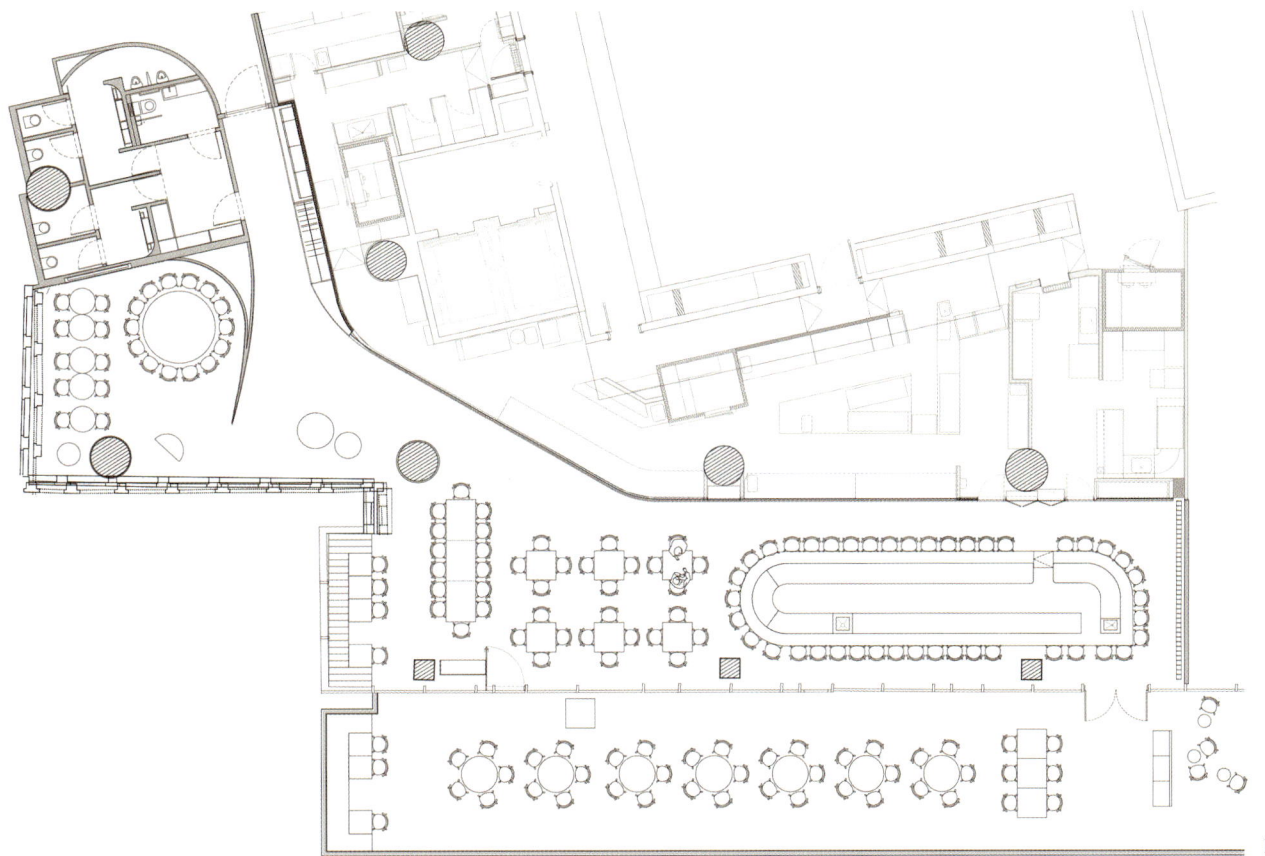

Floor Plan

103

The end of the architectural view culminates in the sake display library, constructed of brushed aluminum which defines the end of the space and functions as the backdrop of the restaurant's central bar. The seating experiences are also conceived as broad gestures and comprise mainly of an elongated bar located in the center of the space. The dining tables are designed in the same language as the patina cladding and are made of aluminum plates in two finishes. The private dining hall enjoys an oversized table for communal seating and can be transformed into an even larger piece that can host up to 20 people.

The lighting of the space is mainly indirect and washes the arch from below, while above the bar and the private space the lighting becomes visible in the form of a modernized version of a chandelier, made of aluminum thin pipes that incorporate little led lights, governing the dining in these areas. The space is essentially monochromatic in shades of light warm grey and draws its inspiration from modern Japanese architecture that is sculpted in concrete. The copper patina and wood of the dining chairs endow the space with accents of color, and the terrazzo grey floor introduces sprinkle of green and recalls European materiality while also referencing local Israeli heritage. A is essentially a delightful expression of a modernist interpretation served on a plate of classic inspiration.

107

Jufka restaurant
Pristina, Kosova

The space is further enriched with small hand drawings with Italian motifs that are drawn on the walls or wooden décor of the restaurant and with the use of hanged greenery that encompass the space.

A restaurant design that highlights on the comfort and joy of life is exactly what the design of Jufka Restaurant represents. Intertwined with the use of traditional elements, the lightness of the space is an identifiable characteristic of this design. The motif behind the design stands solely on the fullness of traditional Italian restaurants, where the eating seems to be celebrated as an activity happily shared with loved ones. This is reflected on not only the food being served but also the atmosphere created by the design of the restaurant itself. This same feeling is brought forth through the design of the Jufka Restaurant.

An important part of the interior philosophy of Italian restaurants is that the cuisine is matched by a beautiful, clean, light atmospheric interior. Inspired by elements of Italian interiors, the feeling of this designed space is created through the use of three main materials translated into interior motives that Kosovars can also relate to. The play of materials includes mosaic tiles, textured plaster, and the use of wood. The lightness of the space is determined by the light earth tones of the interior: light green for the mosaic tiles, white for the texture plastered walls and wood for the furniture.

Project details

▶**Design:** Maden Group
▶**Homepage:** https://madengroup.com
▶**Area:** 510 m²
▶**Location:** Prishtina, Kosovo
▶**Photographs:** Leonit Ibrahimi

About us

In order to offer Kosovo's architecture a unique studio, that gives comfort, spreads friendly vibes and works with creativity, we decided to open Maden Group in 2009. We always try to integrate new elements according to the evolution of technology and materials, giving different dimensions and colors to our creations but never losing our identity. Our main attention is the environment and the atmosphere where we work in. In pursuance of refreshing the work we do, sometimes we transform our workplace to avoid monotony and set up a creative surrounding. This has given amazing results in our work and also enhanced fondness and friendship in our team.

In addition to the materials, the space is arranged around a central round space that along with the planted tree

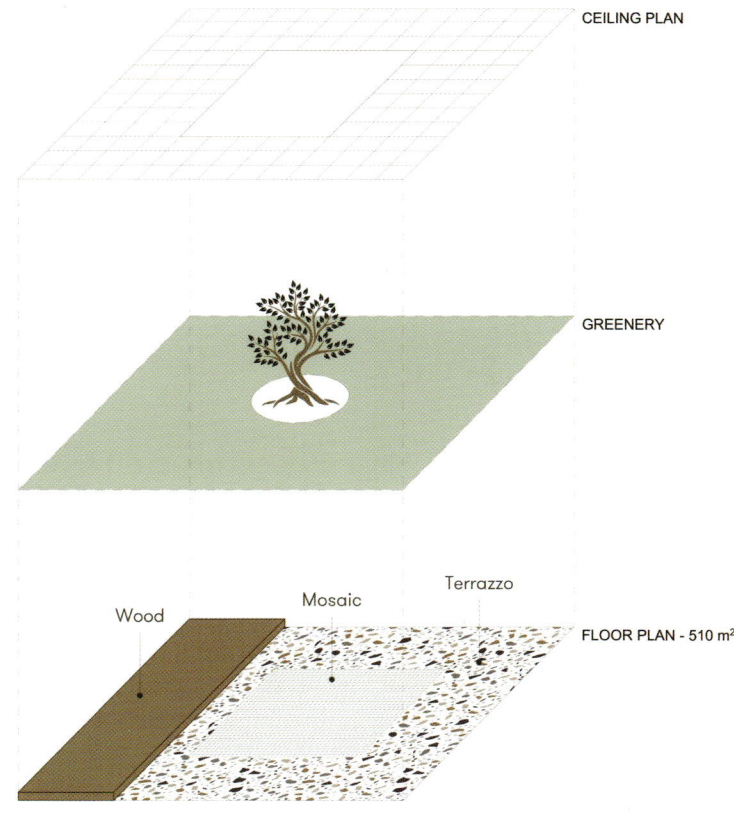

Design Concept

CEILING PLAN

GREENERY

Wood   Mosaic   Terrazzo

FLOOR PLAN - 510 m²

becomes an identifying element of the restaurant as well.

Floor Plan

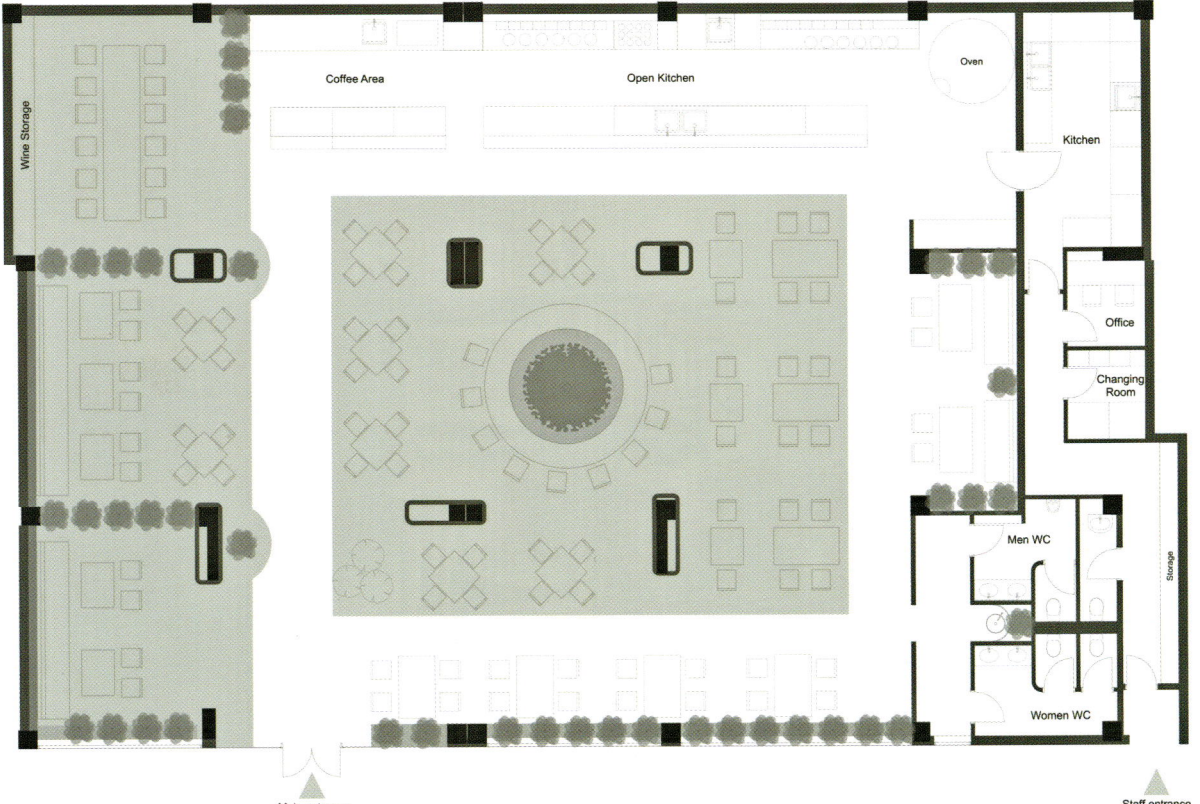

115

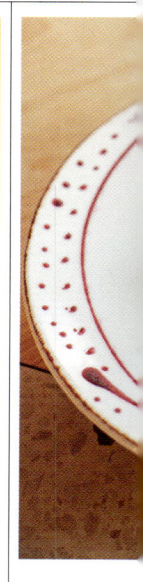

The kitchen is kept opened and inviting to the costumers and treated as part of the joyful experience of eating, where even the preparing of the food remains transparent, as in other traditional Italian restaurants. Everything happens openly; the guests can easily enjoy their food as they watch the process of making pizza and pasta. Another identifying element is the arch. Although largely expanded, arches are used to create divisions within the space and offer also more intimate spaces for smaller gatherings and events.

The space is further enriched with small hand drawings with Italian motifs that are drawn on the walls or wooden décor of the restaurant and with the use of hanged greenery that encompass the space. All of these elements combine harmoniously to create a continual spaciousness and clean and fresh appearance of the restaurant.

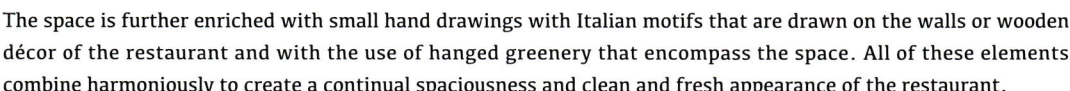

Hokkaidon: 18 Taikoo Shing Road, Taikoo Shing, Hong Kong / +852 2577 0828

Hokkaidon
Taikoo Shing, Hong Kong
Hokkaidon japanese restaurant

A unique take on a traditional motif, hokkaidon's graphic expression is a disruption of the tranquil pattern of seigai-ha, the instantly recognizable blue rolling waves. energy of the sea-to-table dishes unravels on the murals as an array of eclectic seafood break through the blue and white boundaries of the ocean, adding renewed vigor to this piece of japanese heritage. rice, the foundation of japanese food and culture, is at the core of the logo, creating a pure and distinct brand that reflects the essence of chirashi.

This Japanese restaurant is in Cityplaza, specializing in 'donburi', meaning 'bowl of rice dish', with fish, meat or other ingredients served over a bowl of rice. As its name highlighted, they proudly feature fresh seafood coming from Hakkaido on the menu, prepared in the format of chirashi. Seated comfortably, the first thing I noticed was the mural art of indigo waves on the wall, reminding me of the rolling sea where the seafood comes from. The cypress furniture offers a sense of neat and tidiness, and overall, the atmosphere is friendly and pleasant.

Project details

▶**Design:** A work of substance
▶**Homepage:** https://www.aworkofsubstance.com
▶**Area:** 280 m²
▶**Location:** 18 Taikoo Shing Road, Taikoo Shing, Hong Kong
▶**Photographs:** A work of substance

About us

We are a collective of passionate explorers from 15 countries. we exercise the art of forgetting what we know and look to maps of the past, as we scavenge for vibrant stories about the people and cultures that our designs inhabit. With the intention of respecting our intrinsic relationship with the environment, we are dedicated to nourishing and strengthening the communities that we touch, to conscientiously using our planet's resources, and to actively upholding our integrity by telling stories through design.

We believe that design is a labour of love as much as it is a love for labour. in our quest to harness the power of sustainable design to improve lives, we revel in the intense ruminations of the future of our world, in the scrupulous refinement of details, and in the thoughtful exploration of the rituals that mold our ideas into shape. Our team of strategists, editors, graphic designers, architects, interior designers, product designers allow us to shape every single touch point.

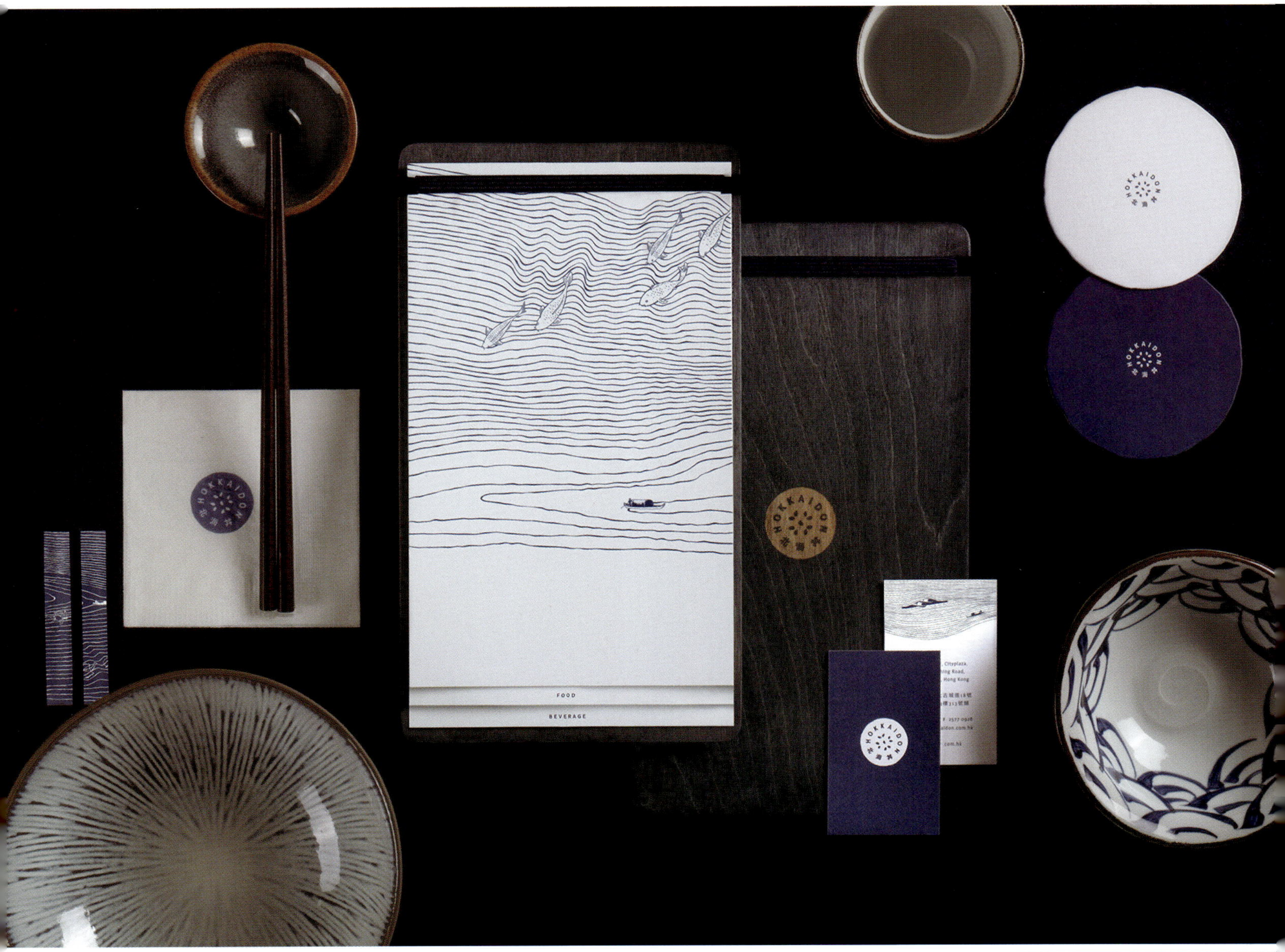
HOKKAIDON
FOOD
BEVERAGE

HOKKAIDON

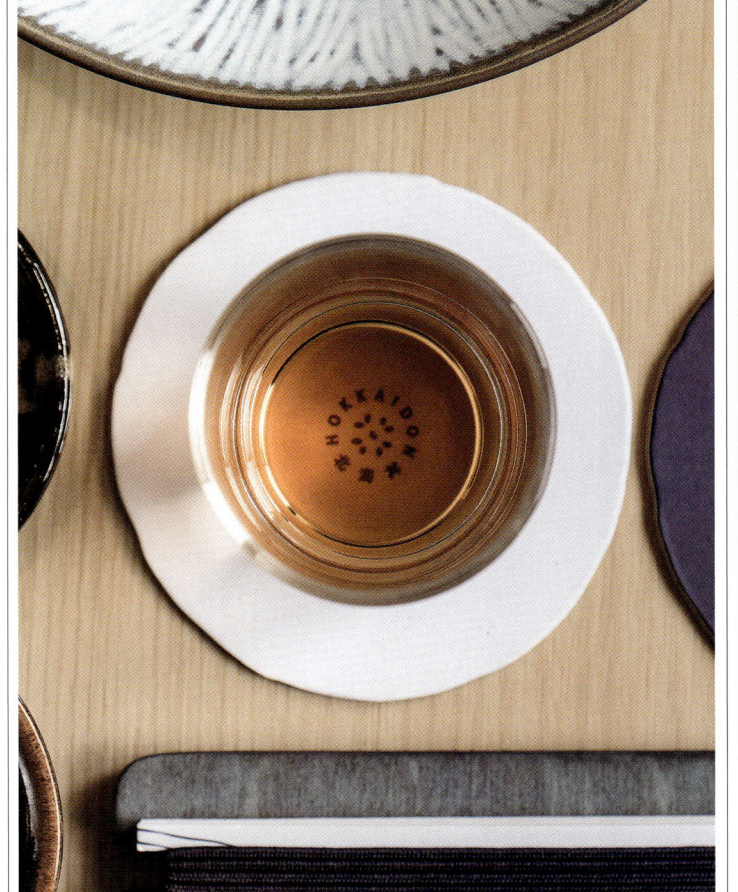

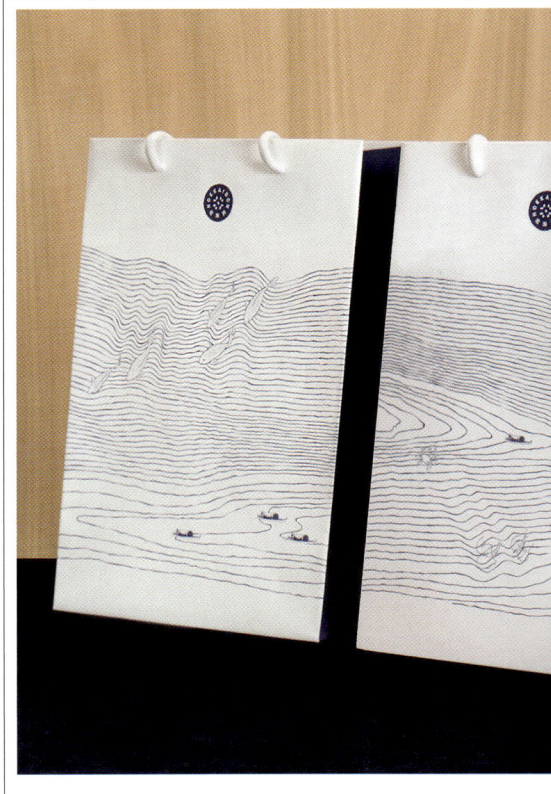

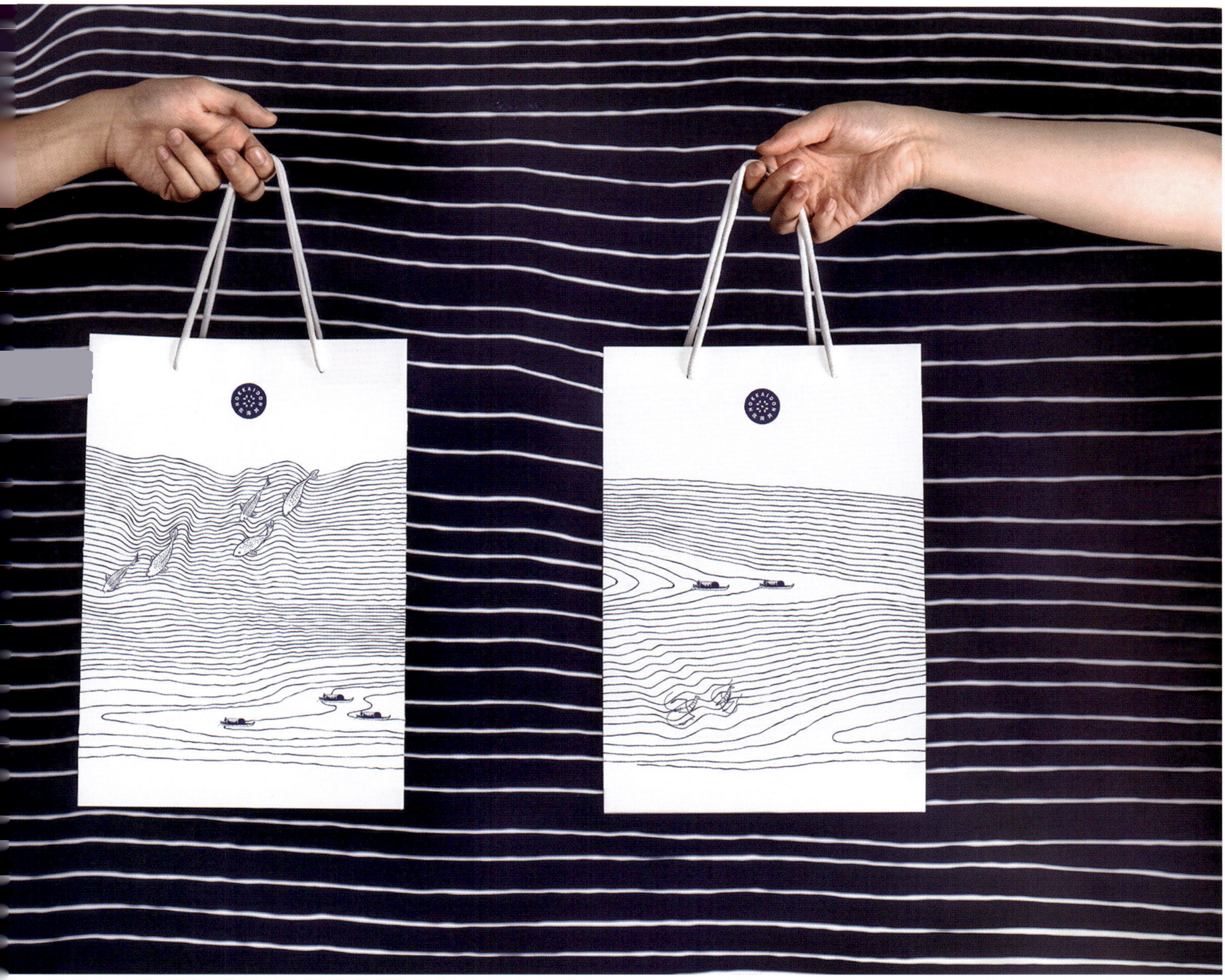

Grain restaurant
Ekaterinburg, Russia

When working on the project, our inspiration was bread and everything that surrounds.

Hokkaidon: Ekaterinburg, Russia / +7 950 209-19-19

Before we started to work on the Grain restaurant project, we had flown to Yekaterinburg. None of the project team members had been there before, so we wanted to immerse ourselves into the context of the city, understand what it looks like from the inside and, of course, get to know its inhabitants. We were very inspired by this trip, fell in love with Yekaterinburg and were imbued with the concept that the Borisikhin brothers planned to implement. We returned to St. Petersburg enthusiastic about making a unique design project, very atmospheric and authentic.

The starting point for the conception was the location - the territory of the former flour mill, where part of the historical building was preserved and restored. The concept of the restaurant, kitchen & bakery, extends this line and the most organic decision on our part was to preserve the continuation of this theme in the interior. When working on the project, our inspiration was bread and everything that surrounds it: wooden boards on which it is cut, linen textiles in which bread is wrapped, and sieves for sifting flour. We used simple and warm materials in the project, which look alive in the space and evoke warm associations with home and childhood.

Project details

- **Design:** DA bureau
- **Homepage:** https://da-bureau.com
- **Area:** 200 m²
- **Location:** Ekaterinburg, Russia
- **Photographs:** Sergey Melnikov

About us

The world driven by the conservative professionals has existed for way too long. Our hierarchyless approach is what forms us. Listening to the passionate and fresh-thinking talents is a crucial part of our culture. And this significantly influences our output and, therefore, the environment and society surrounding us.

We believe that innovative thinking, honesty and passion is a mind-shifting combo that brings good to everything we produce. While actually implementing everything we do is crucial for us.

The final touch was created by unusual flasks with bright blue syrup, reminiscent of the flasks used to soak the rum baba dessert. Thus, we got a concrete shell filled with comfort, warmth, and pleasant sensations. Such a union refers us to the aesthetics of crude industrial elevators, where the very grain is stored, from which the bread begins.

Analysis diagram

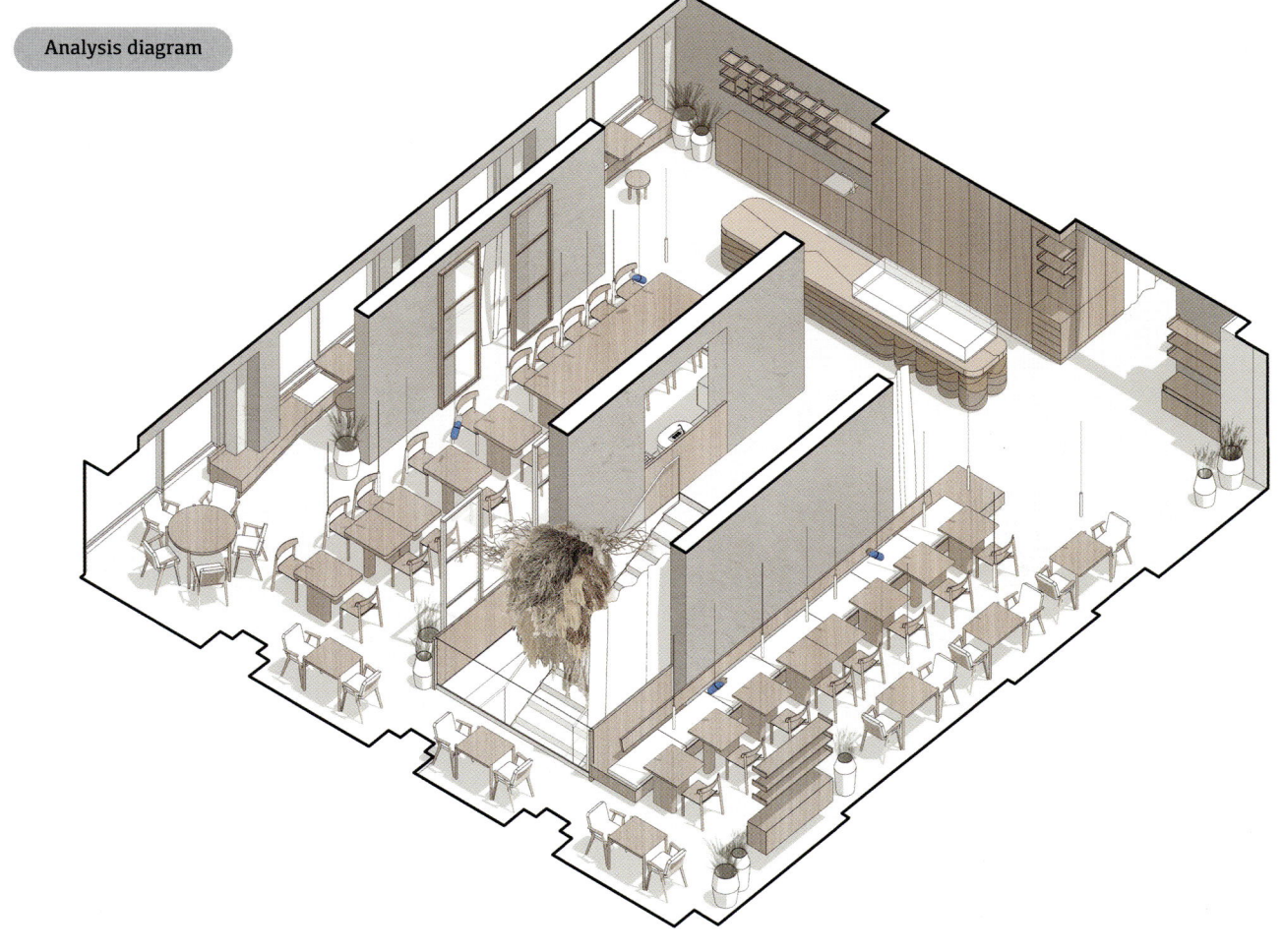

■ When we arrived at the site for the first time, we immediately fell in love with the lighting of the space and its natural, simple beauty. We decided to leave the concrete shell in the same state as we got it from the developer - as clean and open as possible. In addition, we emphasised it with a massive concrete bar, which resembles the shape of a concrete elevator that was previously located in this area.

All the furniture in the space is wooden and functional: the tables look like kitchen boards; erratic partitions dividing the space are like a sieve through which flour is sifted; the scraps of fabric on the ceiling and the curtains refer to the flour-soaked linen in which the bread is wrapped.

143

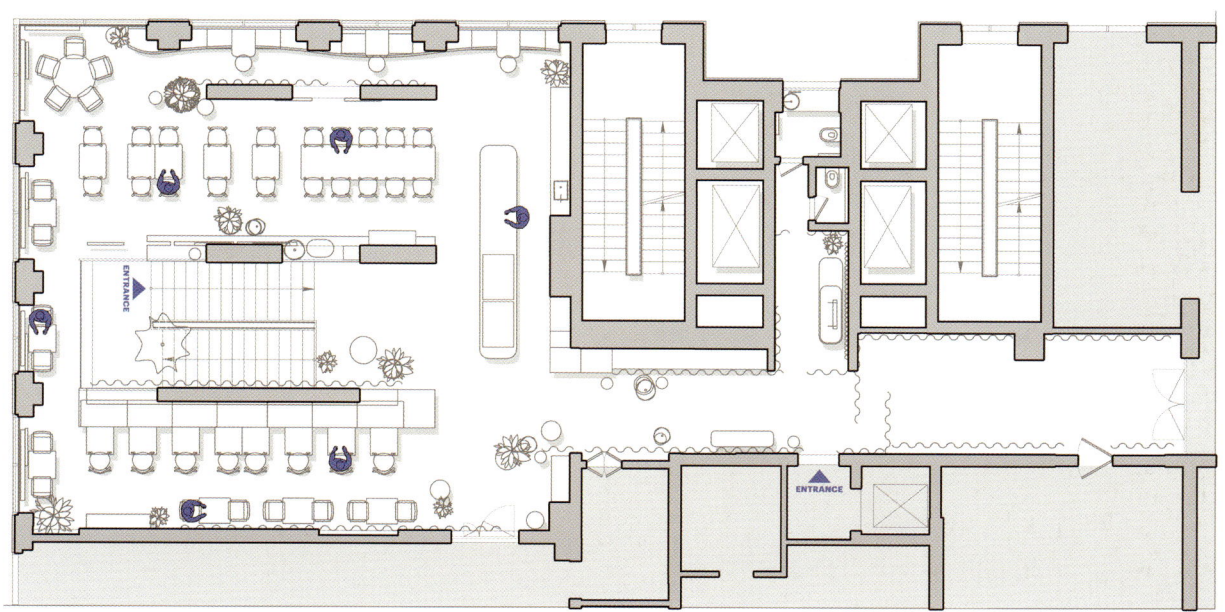

Floor Plan

Tre Dining
Ho Chi Minh City, Vietnam

A harmonious incorporation of tradition and modern elements into our space, sips, and dine to create the quintessential outing experience.

To Vietnamese, 'tre' refers to the bamboo tree, one of the national cultural identities honouring vitality, durability and resilience. To the owners of Tre Dining, 'tre' represents Vietnam's beauty and rich culture. By using bamboos as the primary material for Tre Dining's interior and exterior, the restaurant gracefully reminds diners of the country's traditions and identity.

The concept of Tre Dining is interesting, to say the least. Throughout the space, diners will be surrounded by reminders of Vietnamese tradition, and be served fusion cuisine. In other words, Tre is inducing an exciting fusion of many parts of the world through food and design. Though challenging to achieve as it may sound, this combination was executed to perfection by the brilliant team at Tre. It would be a huge mistake not to mention the play of artificial light in Tre. The restaurant sees indirect lighting features in many spaces to induce an alluring and comfortable ambience. Enabling the moody atmosphere is a series of Unios' lighting solutions. Famous for their high-quality lighting and timeless design, Unios' luminaires assist with wayfinding, class elevation and mood creation. Upon entering Tre, customers are immediately welcomed by a bamboo-filled entrance hallway. Following the stairs, lighted by the hidden Eclipse Linear Lights, to the upper level, Tre's beautiful and purposeful dining space opens up charmingly. The place's sections are separated by curved walls made of bamboo. At the foot of the walls sees the Aeon Flex. The luminaire casts indirect light upward from behind a concrete ledge to accentuate the wall's texture. Alongside the Aeon Flex, the Eclipse Linear Lights were also hidden behind ledges to provide indirect illuminance on multiple linear bamboo walls.

Project details

- **Branding Identity Design:** InSpace Creative
- **Homepage:** http://www.inspace-creative.co
- **Architect Design:** Nguyen Quoc Long & Partners
- **Illustration & Typeface Design:** Nhi Tuong
- **Area:** 290 m²
- **Location:** Ho Chi Minh City, Vietnam
- **Photographs:** TRE | Unios Vietnam | Valor Studio

About us

Dots unite to create lines. Lines splash space into surfaces. Surfaces together build shapes. Shape is where things are born. Inspired by that, inspace would like to be your first dots, together with you 'mold' your brand.

Shapes, bring theme to life, and introduce them to this universe. To us, every brand is born with specific mission, Vision and value.

Contacts e-mail: welcome@inspace-creative.co / Call us: +84 905 66 99 05

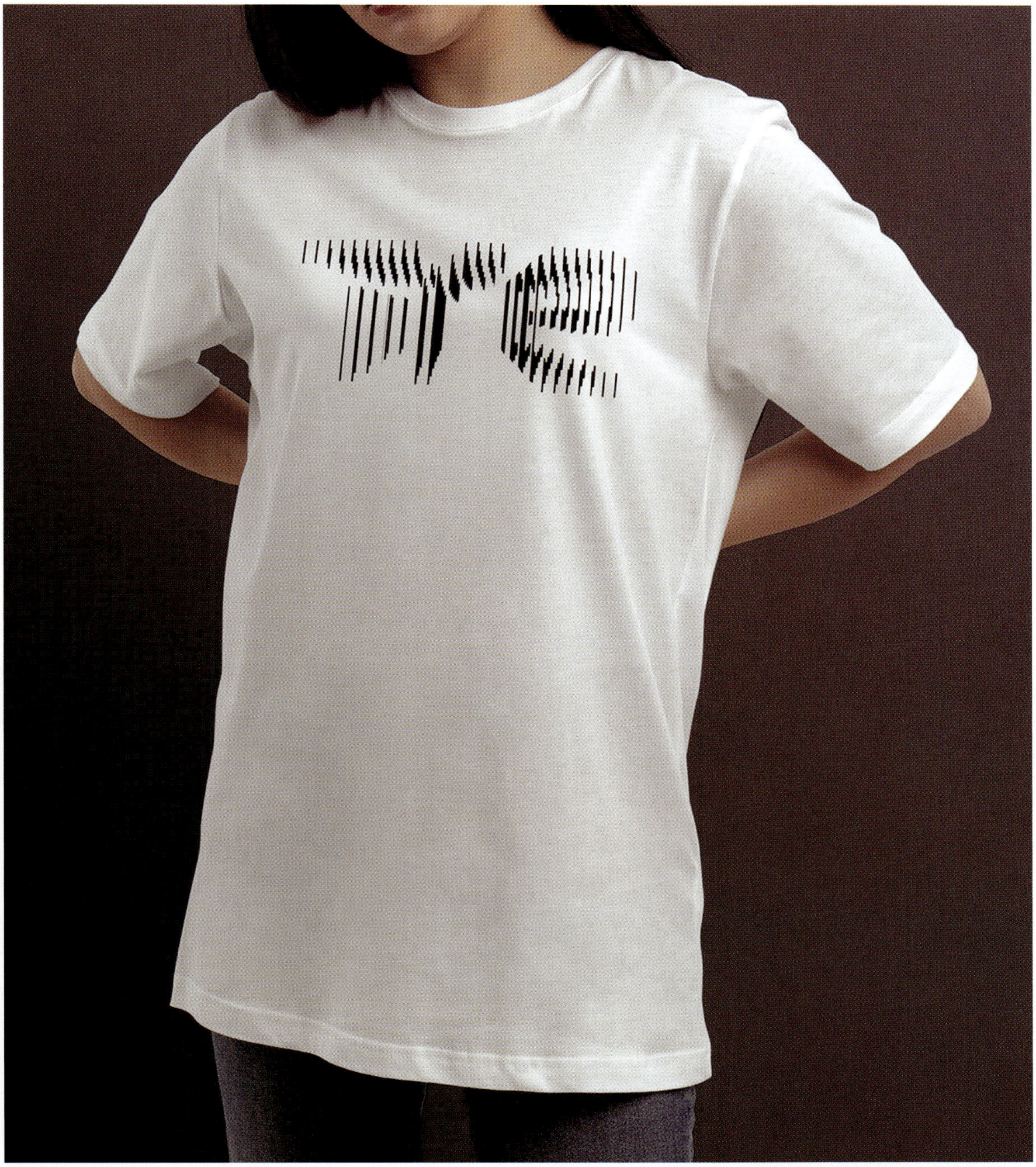

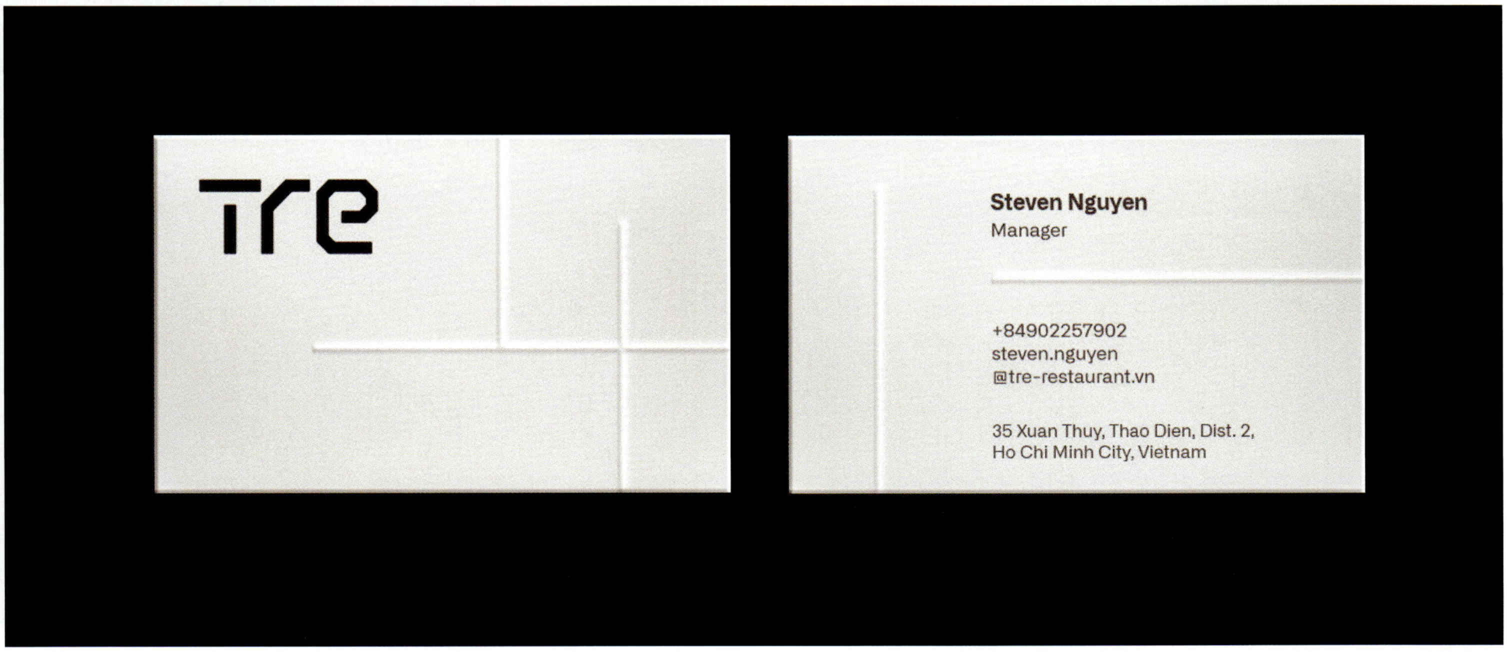
Steven Nguyen
Manager
+84902257902
steven.nguyen
@tre-restaurant.vn
35 Xuan Thuy, Thao Dien, Dist. 2,
Ho Chi Minh City, Vietnam

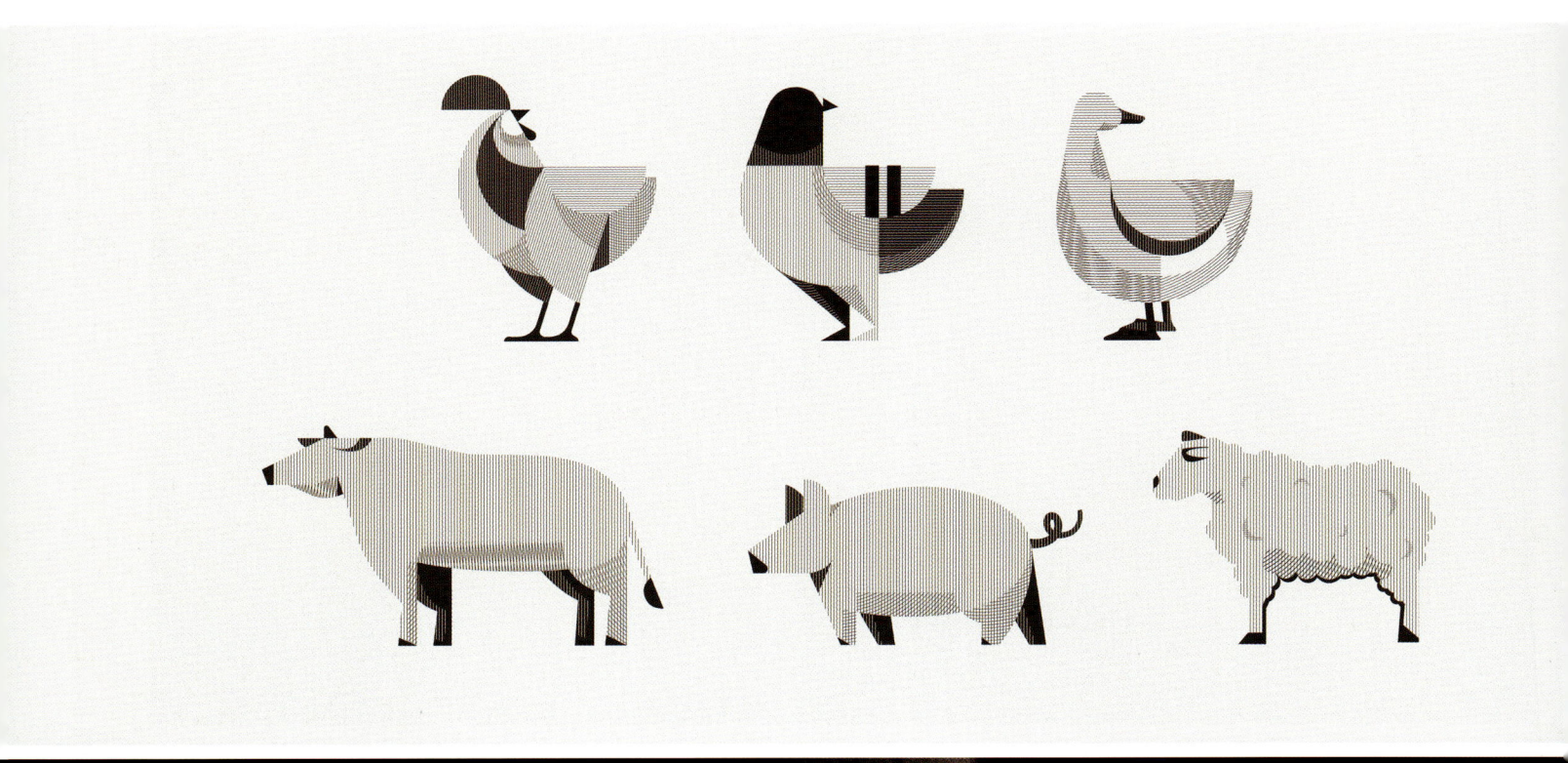

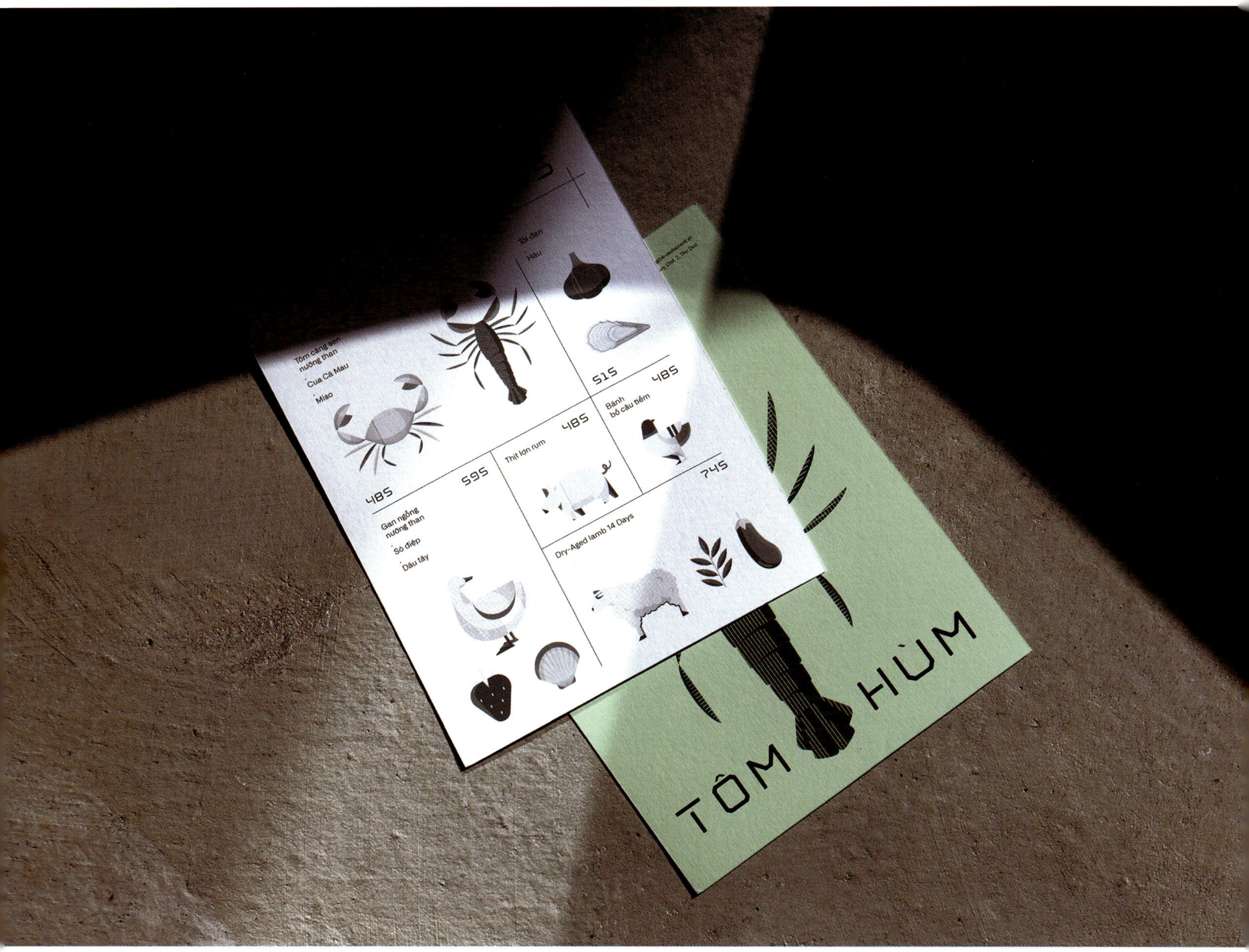
Tôm càng son nướng than
Cua Cà Mau
Miso
Tỏi đen
Hàu
515
485
Bánh bồ câu tiềm
485
Thịt lợn rum
595
485
Gan ngỗng nướng than
Sò điệp
Dâu tây
745
Dry-Aged lamb 14 Days
TÔM HÙM

The graphic pattern identity is built based on the **T** Chim" roof structure of Tre. Tre is a Typeface designed based on the geometrical lines and properties of Bamboo, the curved corners of the characters are rounded to show Bamboo's elasticity and flexibility, firm but still supple. This typeface is used as a Display Font, to represent Tre's specially designed content, creating an impressive and consistent for the identity of Tre. The illustrations are also built on geometric lines, dots and shadows to create uniqueness and consistent with the architectural and interior design details at TRE.

153

T.r.e is a combination of 3 elements of cuisine, drinks and unique space. T.r.e in Vietnamese means Bamboo, the inspiration for all creations of the restaurant. It is a symbol of flexibility, endurance and longevity. The food is an interpretation of Vietnamese dishes. It's a well-balanced mix of culturally diverse cooking techniques, sourcing quality ingredients, and a balance of traditional Vietnamese features, under a "Bird's Nest" made of bamboo - a unique space with "interference" between contemporary and Vietnamese culture.

Offering a low-profile extrusion and high-quality luminance, the Eclipse Linear Light is installed under the outdoor seating area, at the foot of the menu counter, atop the open kitchen and behind the bar for wayfinding and aesthetic alignment. Casambi controlled, the luminaire offers flexibility in colour illumination at the bar, allowing different moods to be channelled. Assisting Tre Dining's general lighting tasks is a combination of the Titanium Starlight and Particle. While the recessed Particle delivers a comfortable lighting environment with its low-glare and deep recessed body, the suspended Titanium Starlight offers a modern and timeless touch to the space.

Tre Dining's desired representation of Vietnam's culture truly comes to life with the addition of arts and decors. Shining light on these is the versatile illuminance of Unios' Scope Spotlight. Comes with flexible beam angles and excellent accentuation effects, the luminaire is also seen at bar counters to highlight the eatery's focal.

Pukkel Restaurant

Huesca, Spain

The nature of huesca inside this restaurant in spain.

Spanish studio masquespacio designed pukkel, a fine dining healthy food restaurant in the city of huesca, spain. the clients wanted to offer above signature healthy food, a sensorial experience beyond gastronomy so the designers proposed to work with 100% natural materials and integrate nature into the space.

Masquespacio started to search for a connection point between the city of huesca and the healthy gastronomy, based on a salutary lifestyle. ins. in there, they found the reference they were looking for that fit perfectly with the healthy lifestyle concept from pukkel.

Earthy tones of brown, white, and green dominate the space reminding guests of nature while touches of gold are added to give the atmosphere a little bit of sophistication. the main forms in the space are imperfect and mainly organic, they draw a path on the floor reminiscing the feeling of walking through a forest. terracotta is used for the floors, bars and some of the walls with patterns custom-designed by masquespacio for pukkel and rough stucco is applied to several areas as another reference to nature.

As for the layout, circular spaces are recreated around all the paths reminding customers of mountains and allowing them to enjoy pukkel's food and share moments with family and friends away from daily routines. lastly, lush arrangements of plants and flowers that will change depending on each season of the year, decorate the space giving it a fresh feel and bringing more nature into the inside.

Project details

▸**Design:** Masquespacio
▸**Homepage:** https://masquespacio.com
▸**Branding Identity Design:** Igloo
▸**Area:** 140 m²
▸**Location:** del Parque, 3, 22002 Huesca, Spain
▸**Photographs:** Masquespacio, Paola Coiduras

About us

In 2010 Ana Milena Hernández Palacios and Christophe Penasse founded design studio Masquespacio with the aim to do something different ... something that would make us proud and transmit our aim to seek innovation for every project. We have had the luck to work in several countries around the world, discover new cultures and make new friends. Thanks to our innovative approach in search of unique and out of the box experiences for each design, we have been awarded with a lot of relevant international awards. In 2020 we were named 'Young Talent of The Year' by Elle Decoration International Magazine, while in 2019 we were awarded 'Interior Designers of The Year' by the Spanish edition of The New York Times' T Magazine. Previously we also have been awarded with the 'Massimo Dutti New Values' award by Architectural Digest Spain and the 'Wave of the Future' award by Hospitality Design USA, next to a continued international recognition by media specialized in design, fashion and lifestyle trends.

Design and illustration of custom packaging for the "takke away" service of Pukkel, a new restaurant located in the city of Huesca, Spain. Pukkel is a healthy food restaurant that opened its doors at the beginning of 2021. Its philosophy is to work with local and real products, avoiding refined sugars and flours, without unnecessary additives.

At the time of starting the project, the client had many doubts as to whether they would have to offer the "takke away" service before the restaurant itself could open (due to COVID restrictions). For this reason, we decided to show on the packaging itself, through vectorial illustrations, different characteristic elements that customers would find and recognise once they were able to visit the premises. All the illustrations were created based on a geometry that can be found in every corner of the restaurant, such as: the square floor tiles; the circular spaces formed by the tables, benches and chairs; the glassware used for the cocktails; etc. These, and other elements, are represented in the boxes and stickers, which serve as seals for the sugar cane containers, with the aim of giving a more casual air to this home-delivery food service.

It is unusual to find this type of box in a take-away food service. Therefore, we worked on a box that was aesthetically attractive, easy to transport and reusable. We also added a handle to facilitate the transport, if necessary, of several boxes at the same time. For this project we made the packaging for the lunch and breakfast boxes.

Tarjeta de fidelización
@pukkelhuesca
Calle del Parque, 3
22001 · Huesca
634 260 156
pukkel.es
Porque en Pukkel queremos que vuelvas, pero además con premio: UN MENÚ GRATIS
Por cada comida o cena, se sellará uno de los círculos.
Al completar todos podrás disfrutar de tu menú gratis.
Nombre

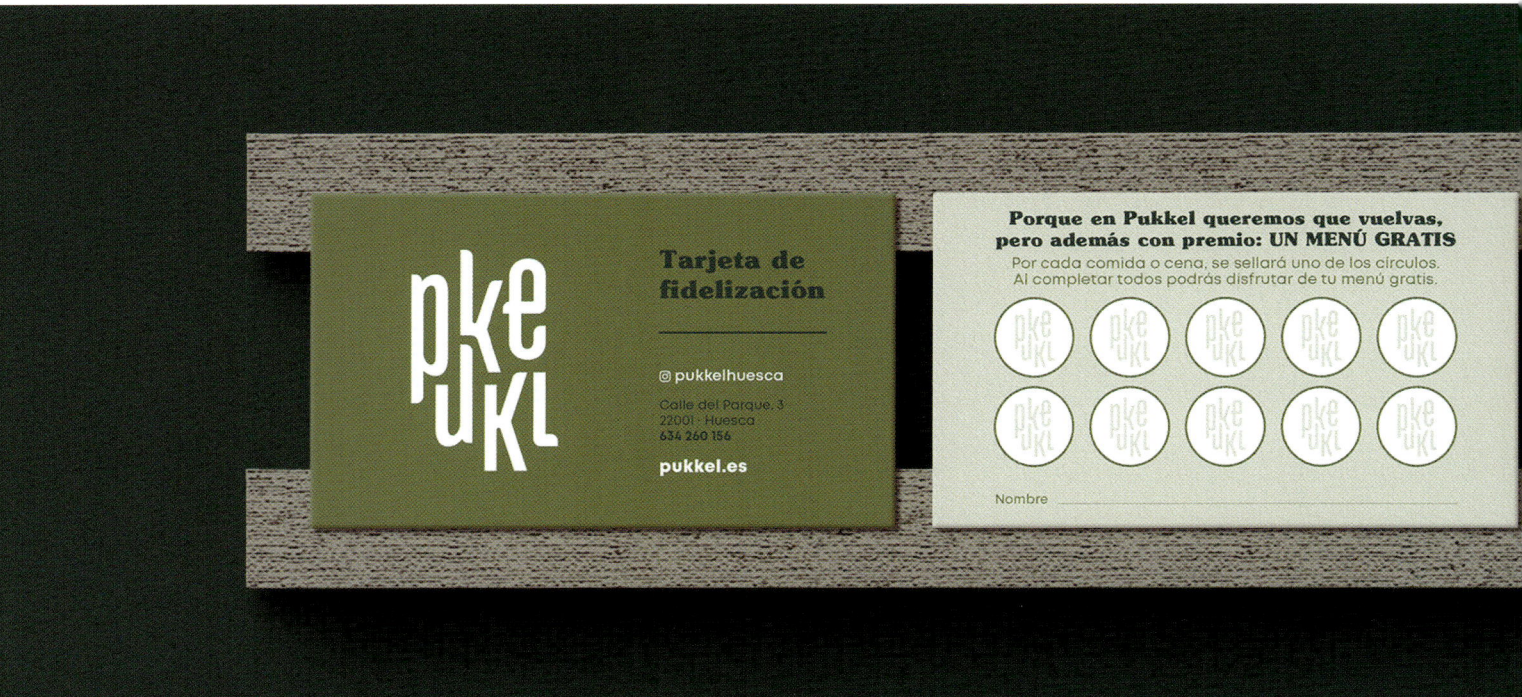

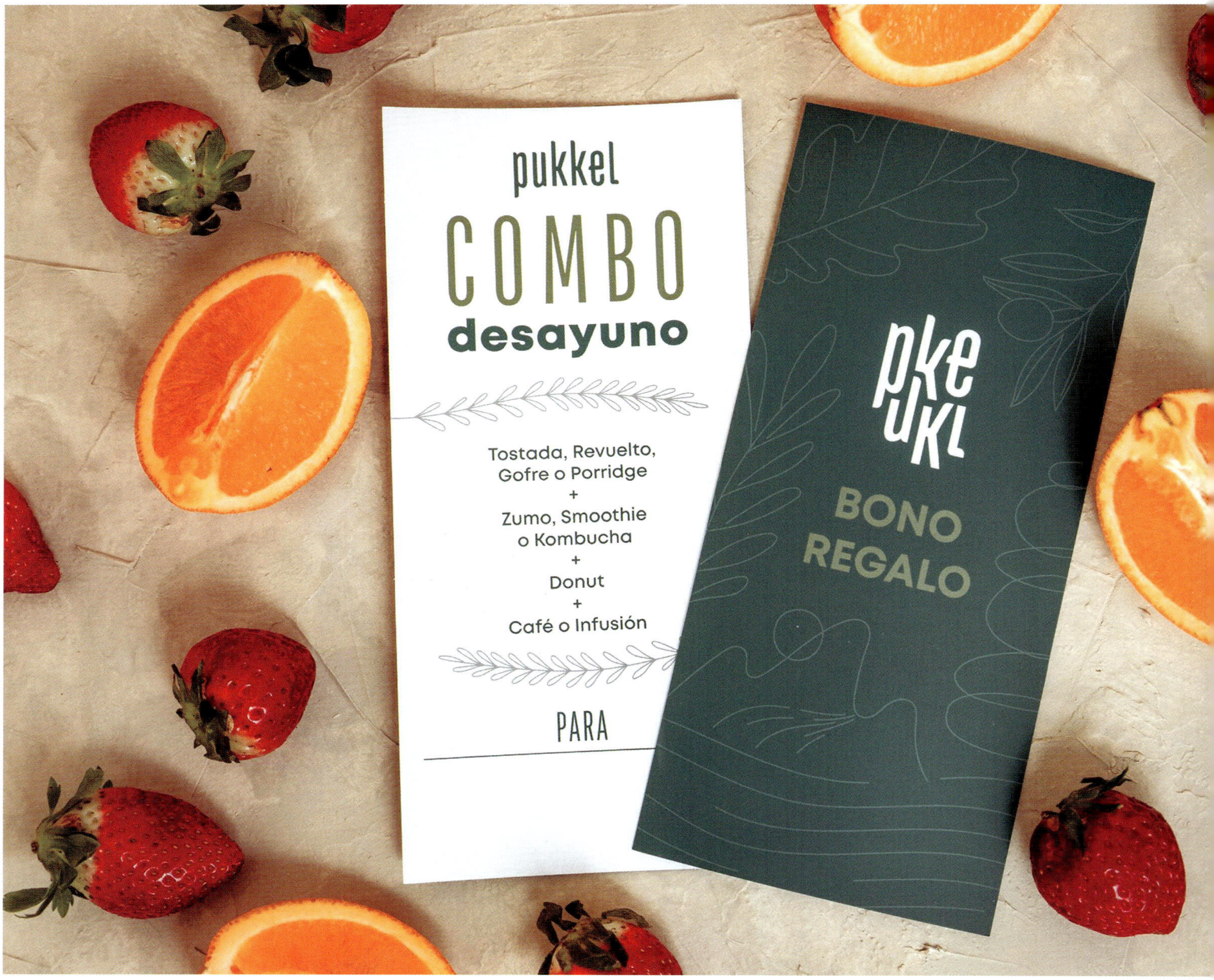
pukkel
COMBO
desayuno
Tostada, Revuelto, Gofre o Porridge
+
Zumo, Smoothie o Kombucha
+
Donut
+
Café o Infusión
PARA
BONO REGALO

The project starts when Jorge and Mikel got the idea to open a restaurant in the city of Huesca with the aim to offer above signature healthy food, a sensorial experience beyond the gastronomy. The beauty of the local mountains and parks in the surroundings was a direct inspiration for this healthy lifestyle concept. Brown and white as well as green tones remind us to the connection with earth of the design.

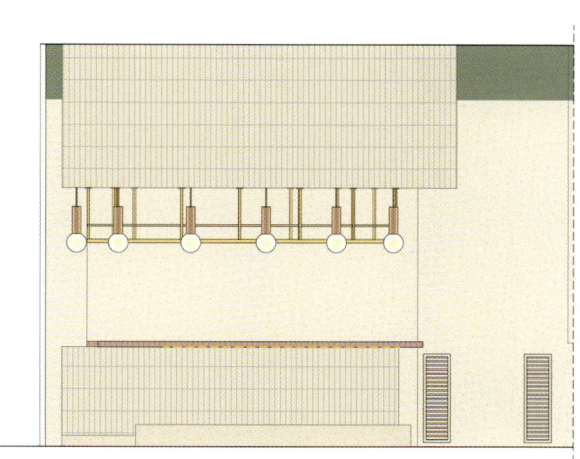

Detail Plan - a

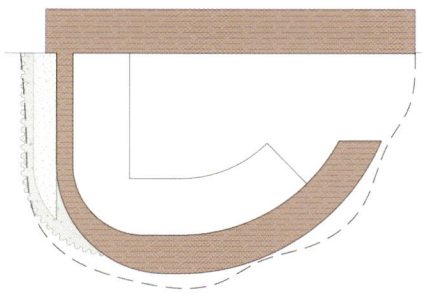

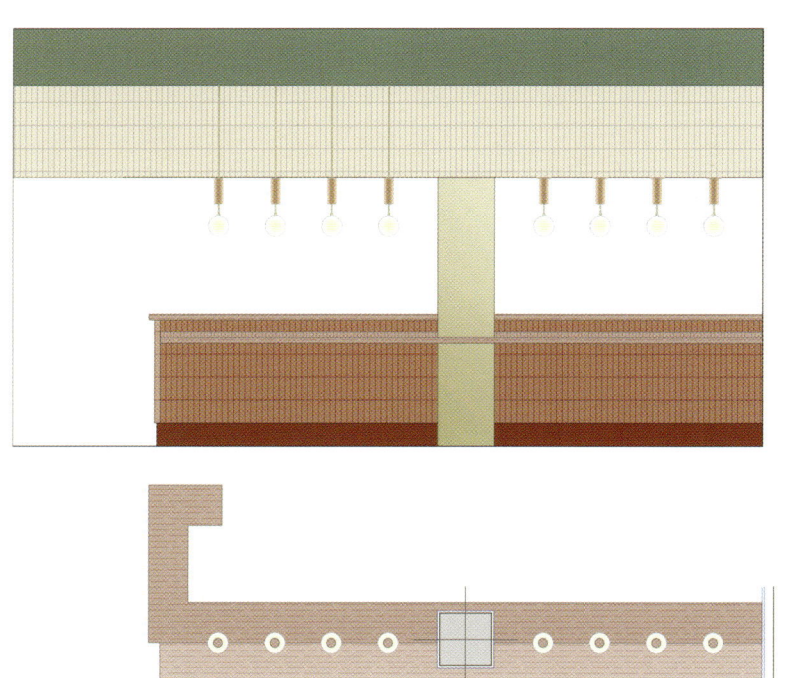
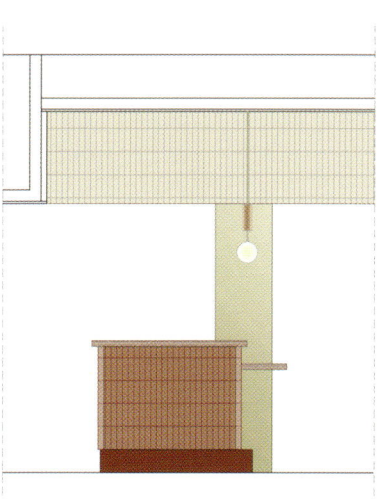

Detail Plan - b

The imperfect forms are mainly organic and draw a path on the floor like if you were walking through the forest. Handmade cladded walls and terracotta floorings, together with the seasonable plants, highlight the 'nature feeling' inside the space as a direct reference to the food concept.

S- B

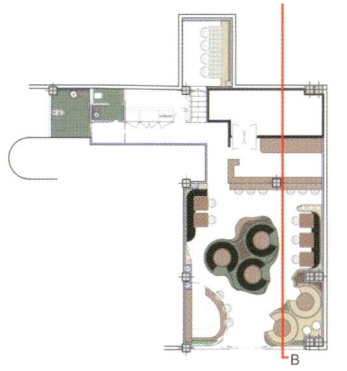
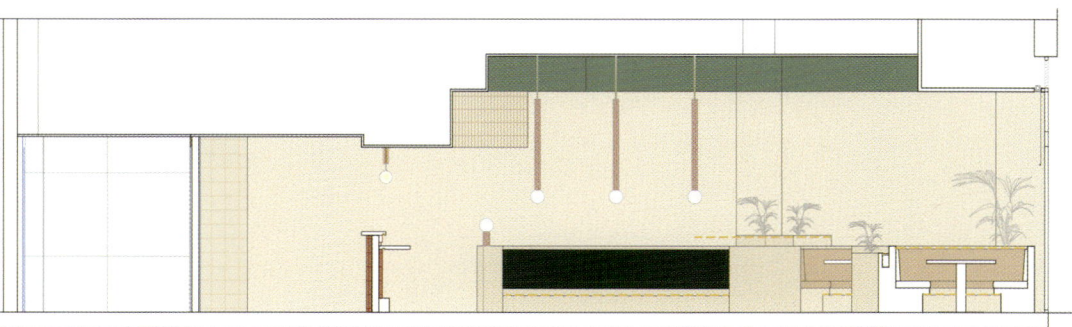

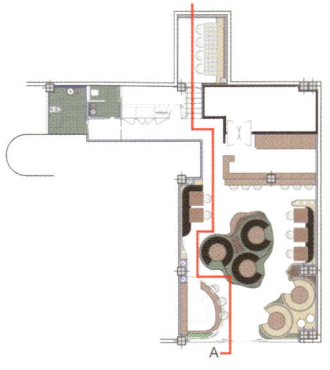
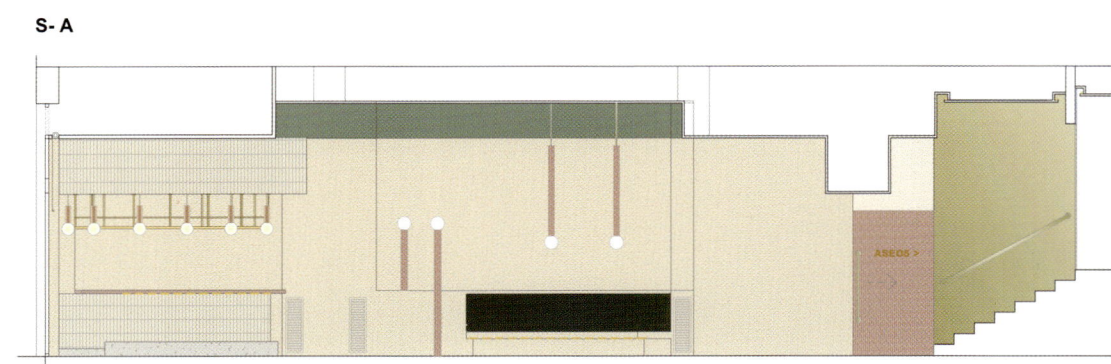

S-A

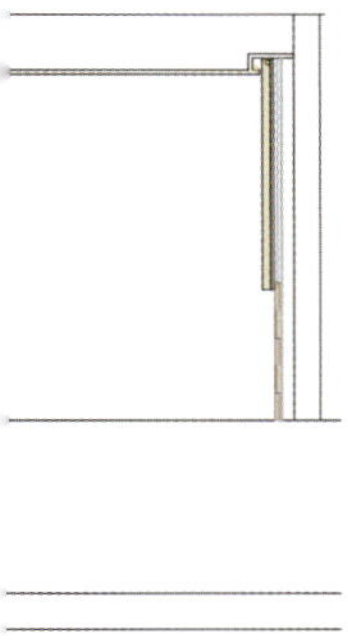

S- C

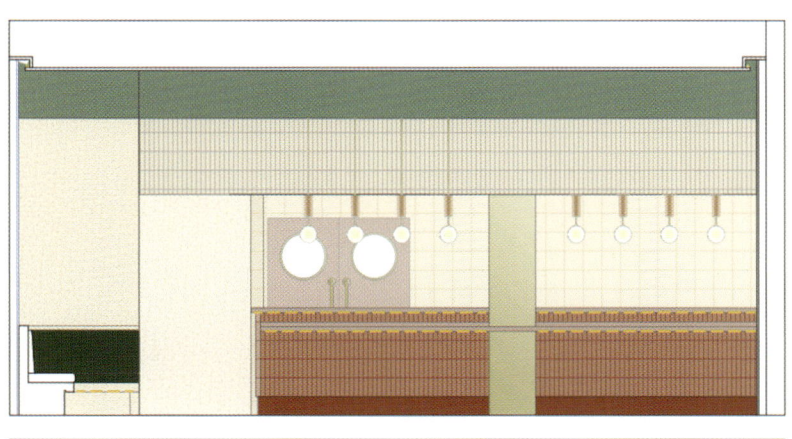

S- D

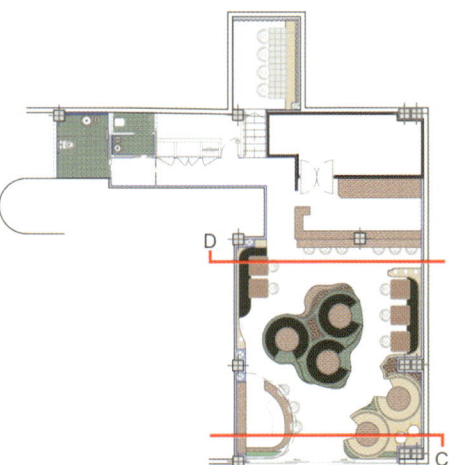

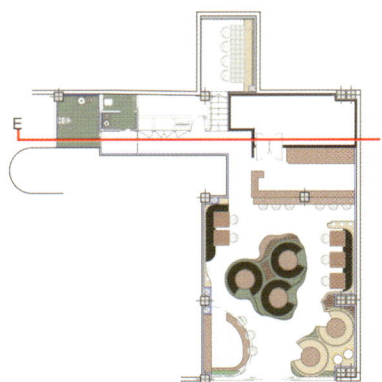

S- E

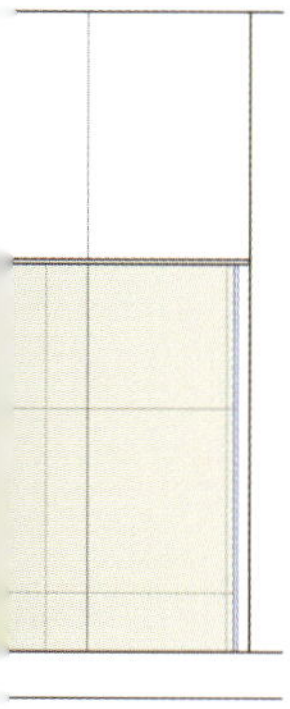

Floor Plan

Mama Manana
Kyiv, Ukraine

Upon our first visit to the site, we decided on a visual design concept of a gradient.

▶Mama Manana: Prorizna Street, Kyiv, Ukraine / +38 (073) 228-11-55

ABOUT "Mama Manana", a new spot of the Georgian restaurant chain, is located on Prorizna Street. It is a historical part of Kyiv just a few steps away from the Golden Gate. The restaurant occupies 470 sq. m. of a building that was erected in the early 20th century.

CONCEPT Mama Manana is a friendly Georgian hostess who welcomes guests at the doorstep of her home and treats them in her living room. Our task was to reflect the generosity of modern Georgian hospitality in the restaurant's interior design.

DESIGN Our space is a four-story building with a staircase connecting the levels. Upon our first visit to the site, we decided on a visual design concept of a gradient. This concept helped us to separate the space: to create areas different in their seating style and functional purpose.

Project details

▶**Design:** Masquespacio
▶**Homepage:** https://www.balbek.com
▶**Area:** 470 m²
▶**Location:** Prorizna Street, Kyiv, Ukraine
▶**Photographs:** Yevhenii Avramenko

About us

Balbek bureau is an award-winning architecture and interior design studio founded by Ukrainian architect Slava Balbek and Borys Dorogov. For 14 years, we have been designing bespoke commercial, corporate and residential spaces.

Comfort, innovation and functionality are the driving forces behind every project we work on. Our approach is to explore the basics and then plunge into details to transform aspirations into ground-breaking environments.

Our work has received multiple international awards and has been published in numerous media outlets worldwide.

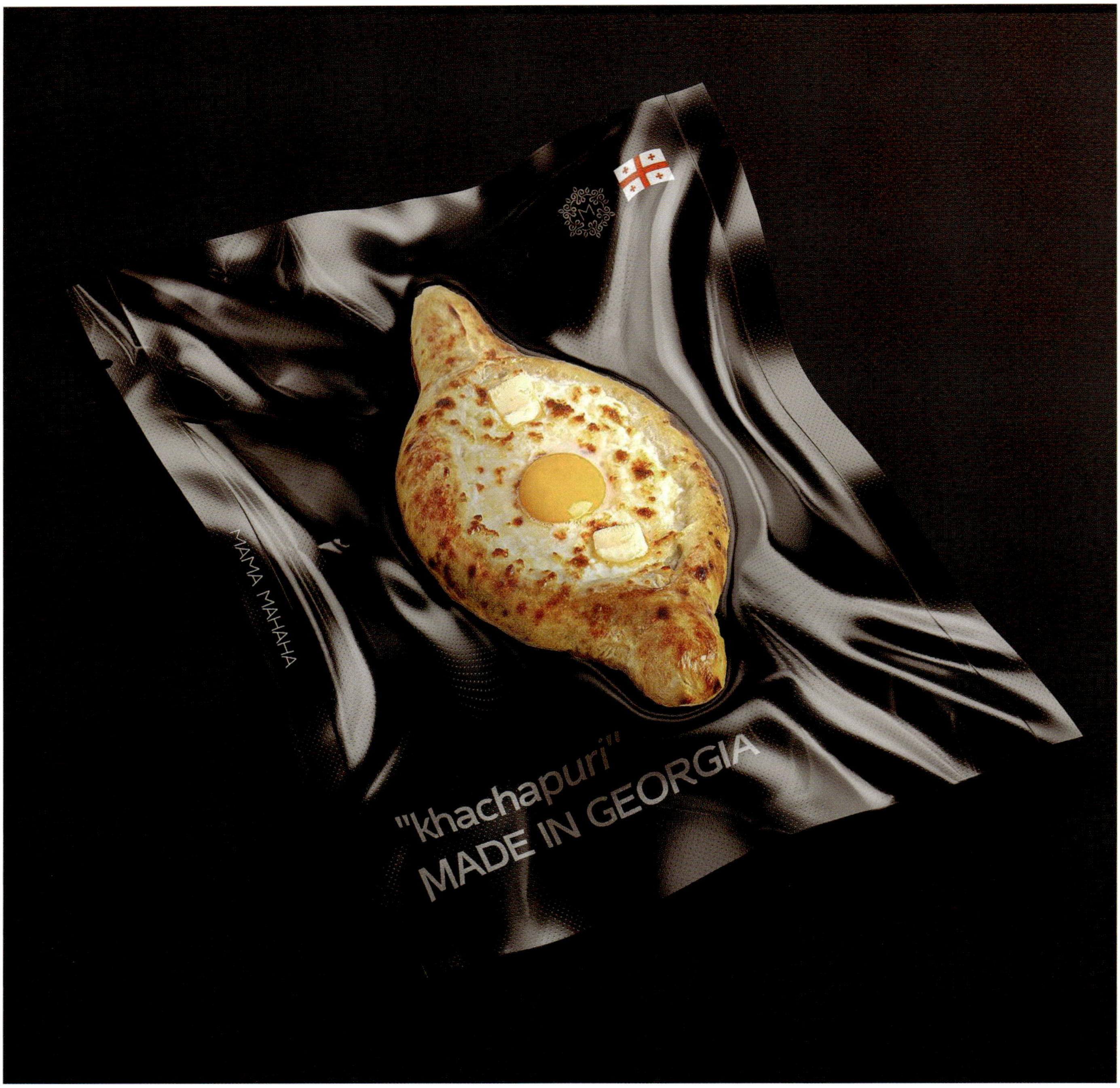
МАМА МАНАНА
"khachapuri"
MADE IN GEORGIA

МАМА МАНАНА
· хачапурна ·
МАМА МАНАНА
· хачапурна ·

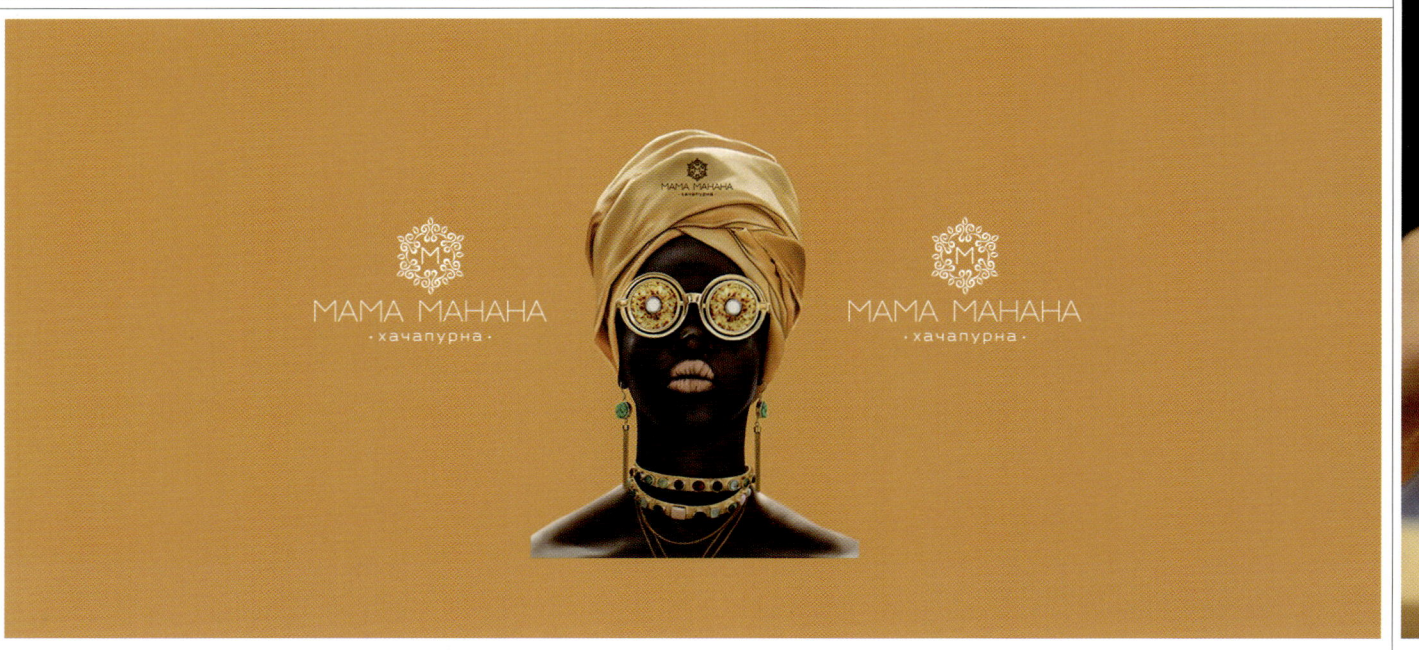

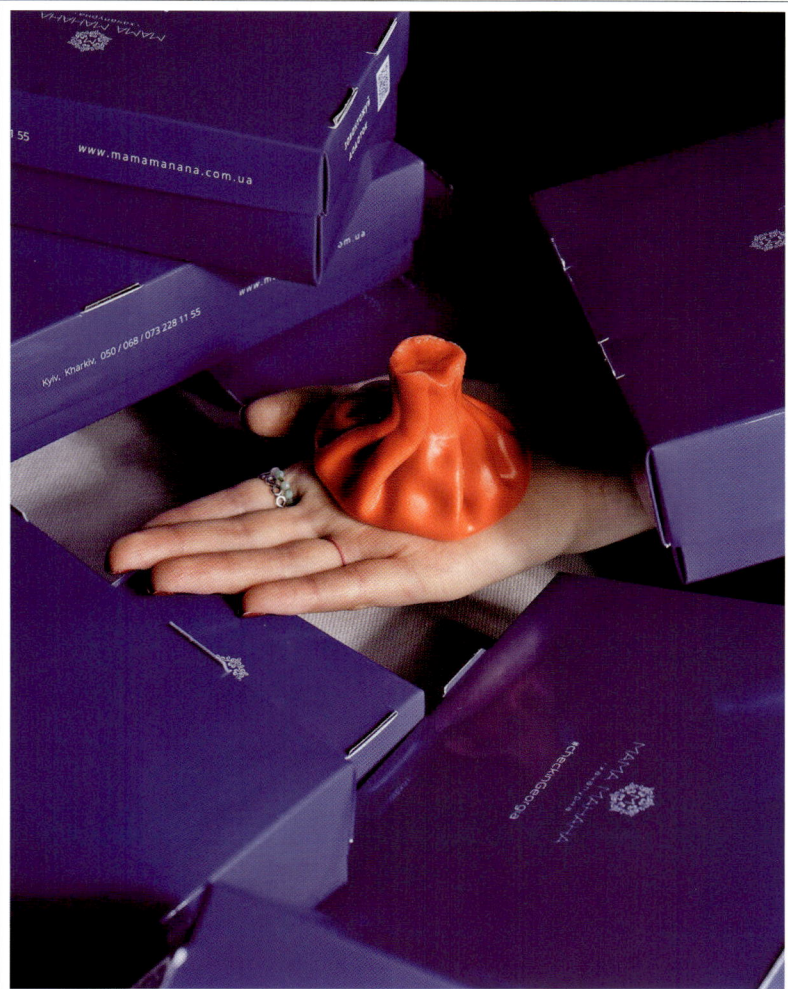

The first level became our starting point. As we ascend, the color scheme becomes lighter and warmer, the materials – softer and more tactile, and the seating – increasingly cozy. When entering "Mama Manana", you find yourself on the first floor of the restaurant – a functionally dynamic space, equipped to seat a large number of people. The hard seating is intended for quick lunches, snacks, and business meetings. Arched windows overlooking the facade of Prorizna Street became the floor's focal architectural element. We preserved their original form and painted in fresh green in the color of the walls. The rest of the materials and shades were selected according to our gradient concept.

Another focal element of this floor is the textured brick walls. Through the play of light and shadow, they create a unique perceptual experience. Some of the brick we used for the flooring, to evoke the feel of a yard or a garden. Also, located on the first floor an open kitchen, allows all guests to experience the process of cooking of Georgian dishes and to immediately enjoy their flavorful aroma. The kitchen area is lined with C.I.Form concrete blocks that contribute to the textural quality of the first floor. Illuminated from the inside, they fill the space with soft warm light.

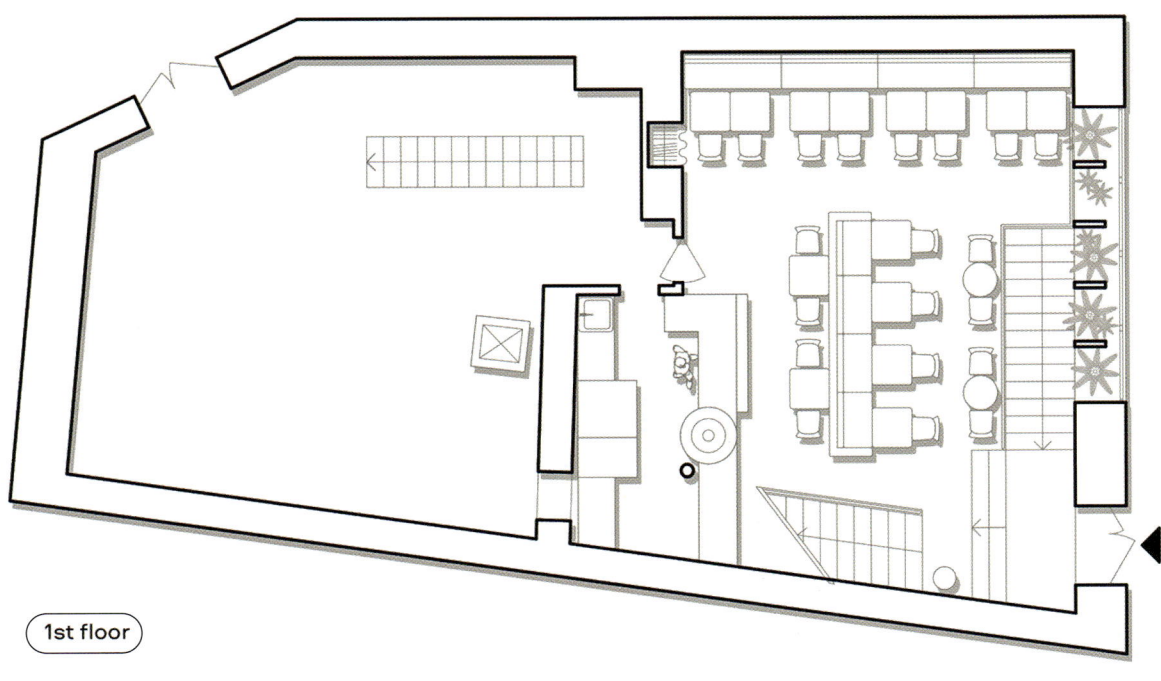

1st floor

Stairs of the black clinker tile lead to the basement where the restroom is located. With a predominantly dark color scheme, the clay washbasin — handmade made by local craftsmen — provides an accentual contrast.

Upon ascending a flight of stairs, we enter into a separated, closed-off area. Called the "banquet room", this space creates a cozy atmosphere for a private feast. Its color scheme differs from the general concept. Metal arches and Georgian ornaments on the walls and ceiling create a unique banquet atmosphere. We designed the ornament based on Georgian motifs, adapting it to the particularities of the space. Armed with only their hands and rulers, the craftsmen spent several weeks putting up this unique ornamentation.

The room ends in a light screen of flax in the shape of an arch, completing the vaulted composition. To close off the area, you need only to shut the heavy velvet curtains; their rich blue color plays to the festive quality of the room. A small dressing room for the guests and an employee-only elevator for serving food is located behind the curtain. As per our gradient concept, we decided to continue some stylistic choices from the first floor to the second. For one, the flooring of the second floor mirrors the ceiling of the first. On this level we chose warmer colors and tactile materials to evoke the comfortable atmosphere of Mama Manana's living room.

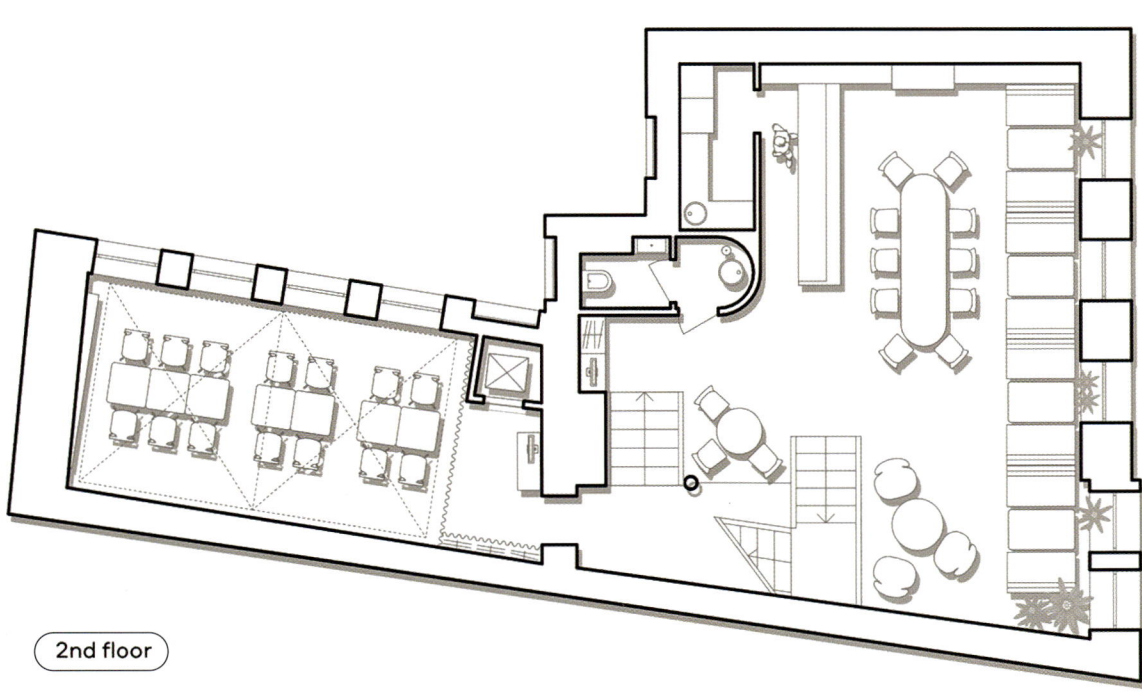

(2nd floor)

■ Along with the window is a soft seating area, which welcomes you to relax and stay awhile. In the middle of the room, we placed a large table and wooden chairs for a cozy family meal by the fireplace. The setting is complemented by a wooden shelf with an extraordinary collection of Georgian wines. The key feature of this floor is our unique deep-green tiled construction with rounded corners. Inside we placed restrooms on one side and bar facilities on the other. It is lined with eight types of tiles, which vary in size and ornament, made uniquely for us by local craftsmen from Detiles Mosaics. Against the green-tiled construction brightly stands our copper bar. It perfectly matches the lamps over the communal table, made of the same material.

The third floor is the highest level and thus the brightest of all the floors. We chose a glass sliding partition rather than a wall to separate the room and the terrace. This increases the floor space and allows for a lot of daylight to enter. In the summertime, this floor transforms into a unified open space. The seating on this floor is a lounge zone with cozy armchairs and sofas. We decided to create two levels of seating on the terrace, which gives the guests sitting near the fencing a wonderful view of the street and the city.

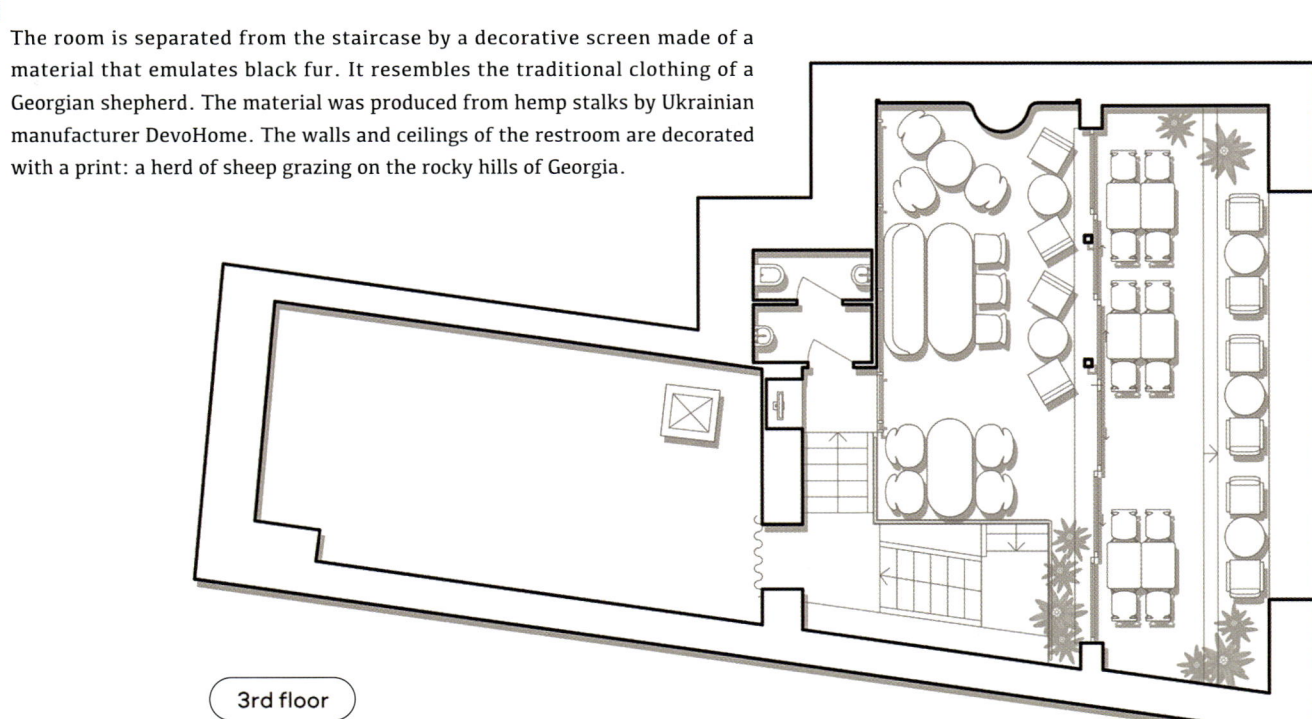

The room is separated from the staircase by a decorative screen made of a material that emulates black fur. It resembles the traditional clothing of a Georgian shepherd. The material was produced from hemp stalks by Ukrainian manufacturer DevoHome. The walls and ceilings of the restroom are decorated with a print: a herd of sheep grazing on the rocky hills of Georgia.

3rd floor

Green Grass Condesa

Condesa, Mexico

Taller David Dana: Located in the Condesa neighborhood on Nuevo León avenue.

The project consists of the remodeling of a two-level restaurant whose main objective was to redefine the image of the place and organize the existing architectural program. The characteristics of the space and its context allowed the development of an environment with perfect natural lighting and views to the outside, giving freshness to the environment and at the same time creating a comfortable atmosphere.

The fusion of natural wood, stones, and vegetation achieved a beautiful neutral approach in modern finishes. The simplicity in its palette of finishes enriches the spaces. In this way, Green Grass Condesa is conceived as an interesting project in terms of its interior design and distribution, where its elements are related both functionally and aesthetically, providing identity and character.

Project details

- **Design:** Taller David Dana
- **Homepage:** https://tallerdaviddana.com
- **Branding Design:** Anagrama(https://www.anagrama.com)
- **Area:** 205 m²
- **Location:** Col. Hipodromo Condesa, Mexico
- **Photographs:** Jaime Navarro

About us

TDDA is a design firm, located in Mexico City. We aspire to create high-quality spaces through a creative interplay of materiality, form, and structure. Driven by a passion for design, TDDA implements innovative ideas through various lines of business.

The workshop was launched in 2013 by David Dana, after having worked several years abroad in Melbourne Australia and San Francisco USA, where he gained considerable experience in the field of architecture.

Under his leadership vision, the workshop has received the following awards: Architecture Masterprize. International Design Award. Association of Interior Architects. Architecture Biennial of Mexico. Open of Mexican Design. Arquine Contest No15.

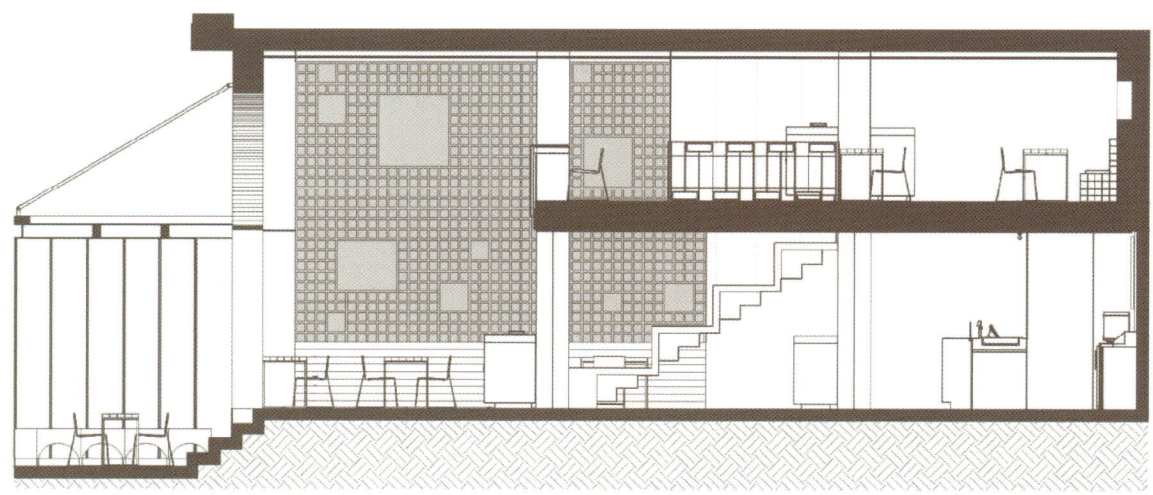
Section

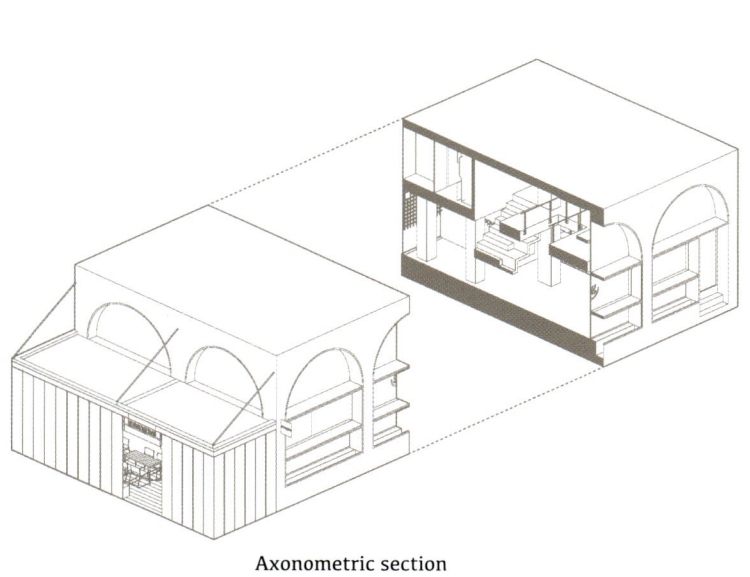

Axonometric section

GreenGrass is a salad shop seeking to change the cultural perception of vegetable-based food in Mexico. They promote the belief that eating is a conscious action, part of a process that nourishes way beyond the physical body. Redesign, update and create a -a visual strategy- for GreenGrass brand that changes the-negative-perception that people have about -healthy food-, reflecting how passionate they are about the quality of each ingredient and keeping in mind the motto "we all desire to become the best version of ourselves" not only in a healthy way of thinking, but also on our emotional well-being and lifestyle.

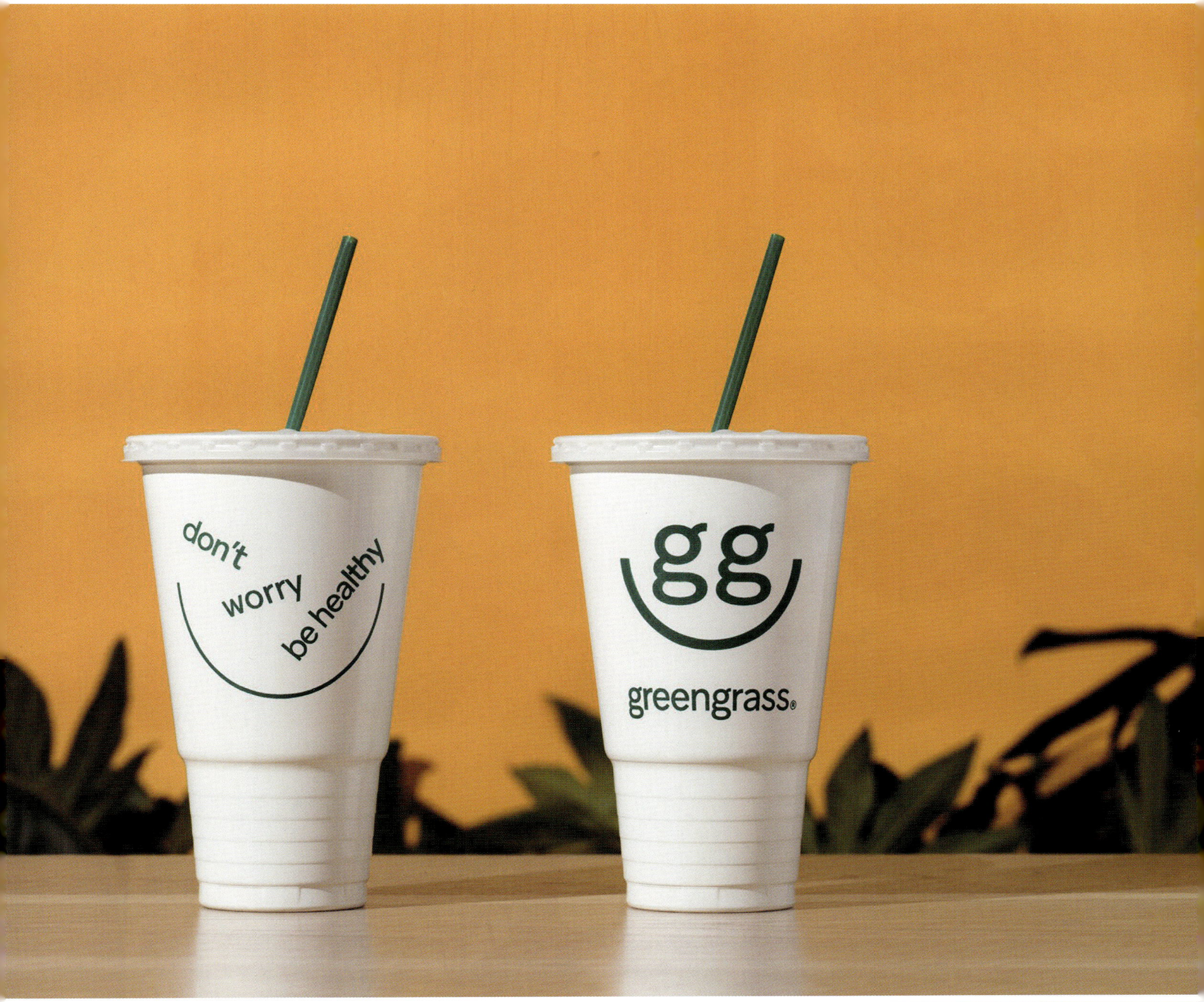
don't
worry
be healthy
gg
greengrass®

greengrass®
agua de fruta
100% natural
greengrass®
gg
agua de fruta
100% natural
480 ml
engrass®

greengrass®
gg
life is better in green
real
food
inside

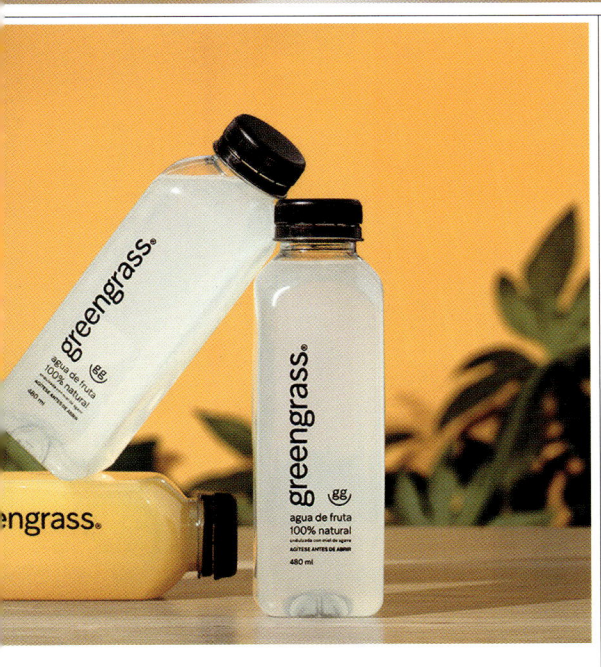

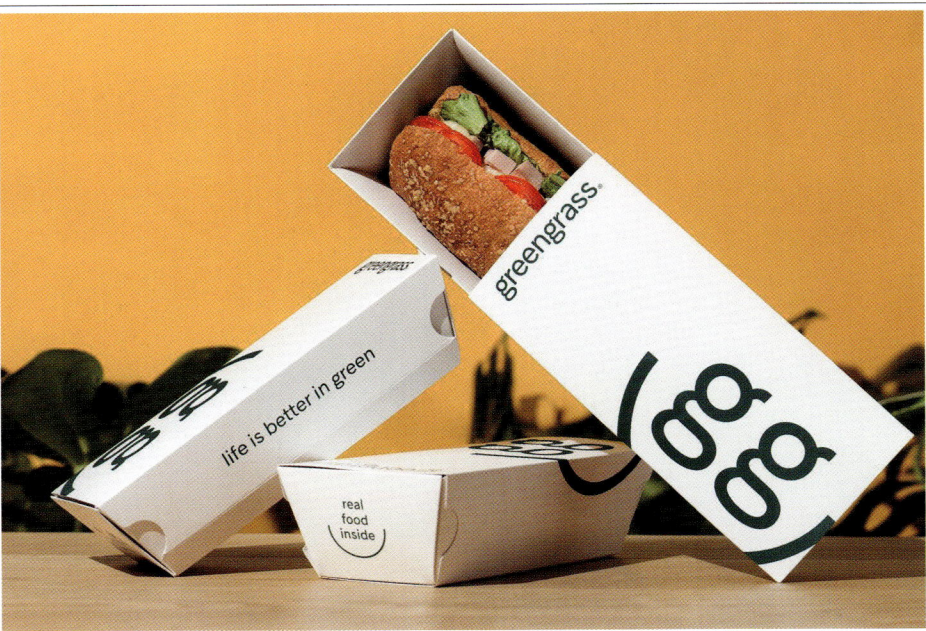

For the GreenGrass redesign, we developed an evolving logo that reinforces the brand values and manages to convey a relaxed and approachable personality.

We also designed an icon complementing the brand identity through a friendly, fun and memorable element, inspired by the spirit of positivity and a smile, which also references the bowl, part of the brand's flagship product: salads.

For the visual strategy we defined the brand system implementing behaviour that communicates the values, talks about the origin and history of the products, and about the quality of the ingredients. This system highlights the power of personalizing and assembling your own salad, communicating healthy food as an accessible, and unpretentious option.

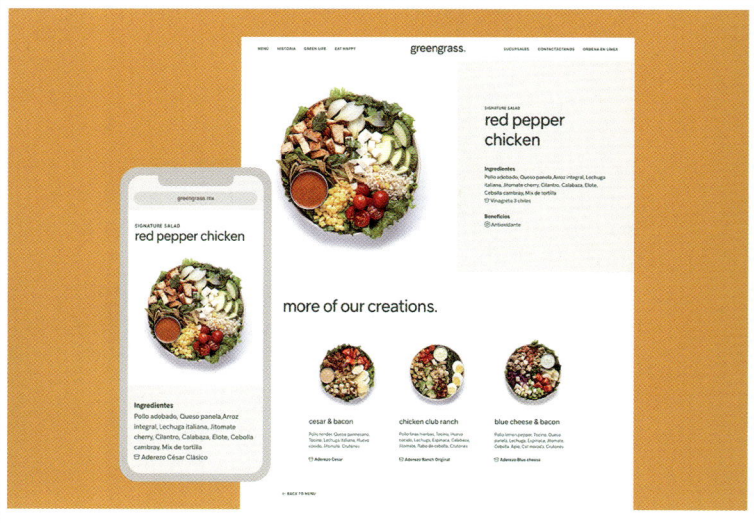

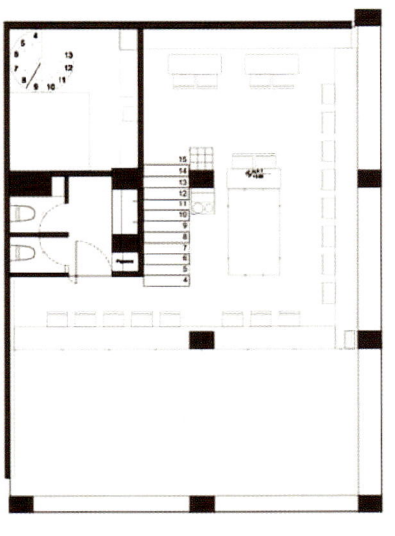

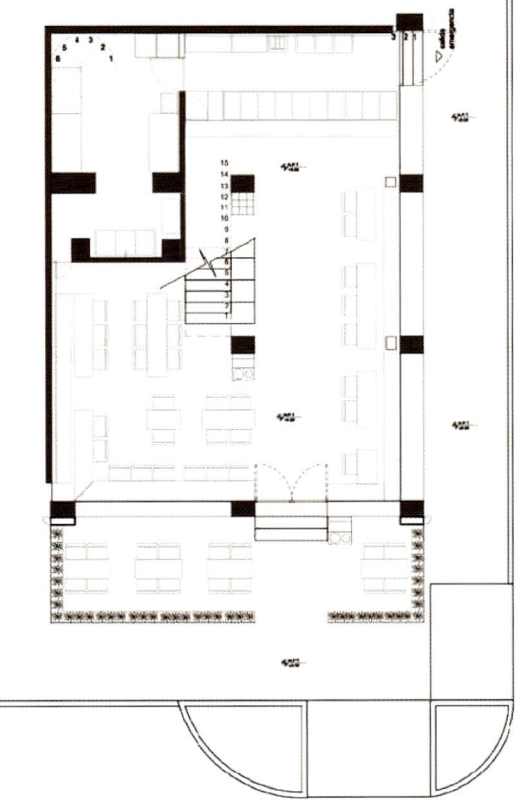

Floor Plan

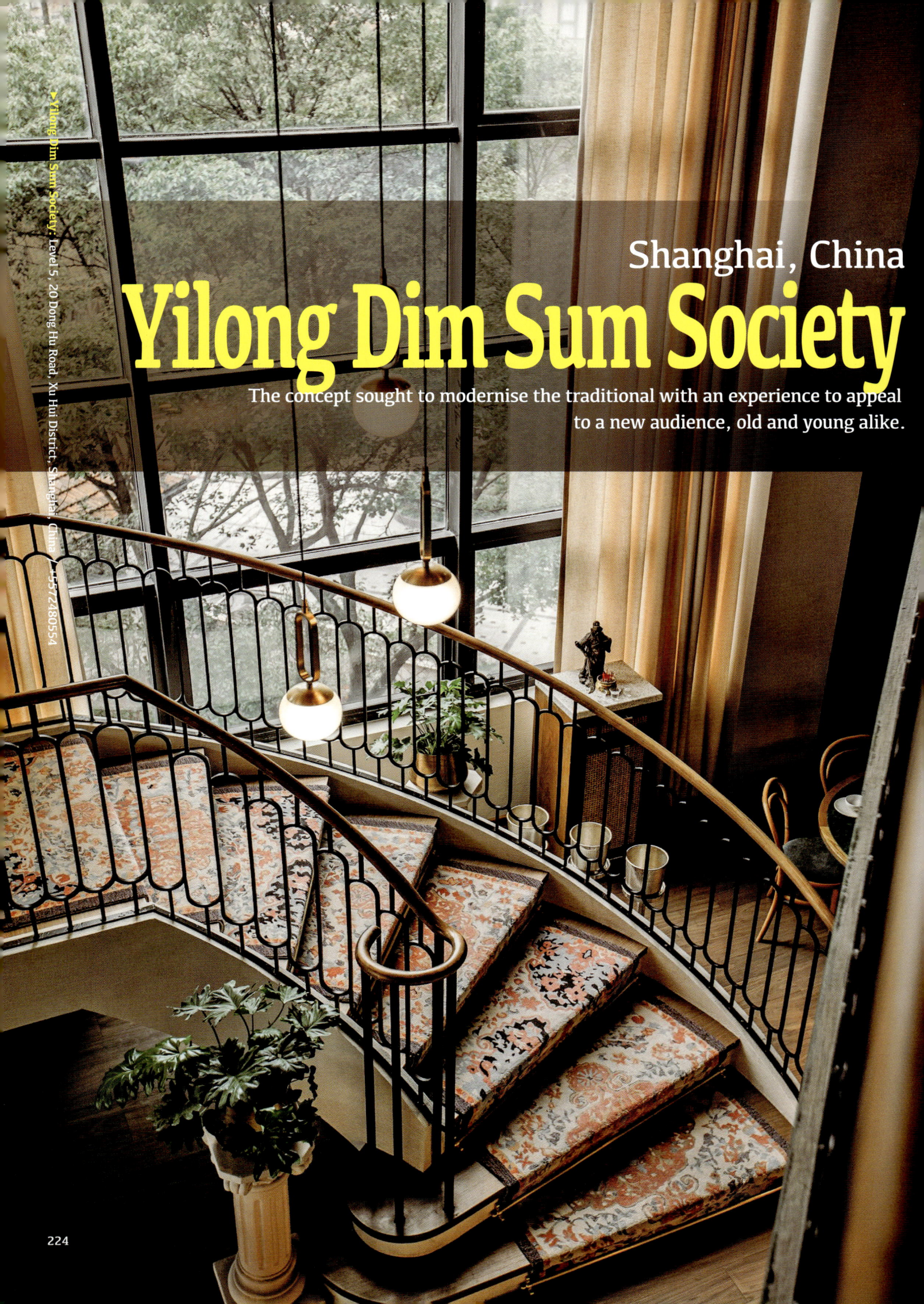

Yilong Dim Sum Society
Shanghai, China

The concept sought to modernise the traditional with an experience to appeal to a new audience, old and young alike.

Project Overview: BrandWorks were approached by the client to create a brand, concept and interior design for a new dim sum, tea and cocktail bar in the heart of Shanghai. Inspired by the golden era of art deco glamour and jet setting between the world wars, Yi Long was modelled off the idea of the 'The Grand Hong Kong Hotel', paying homage to luxurious travel and design of hotels of the 1930s. The project was treated almost like set design as the team dedicated time to watching films like 'In The Mood For Love' and 'The Grand Budapest Hotel' in order to truly immerse themselves in the space. The finishes, details and even styling have all been meticulously crafted to transport diners to the golden age of luxury travel.

Project Brief: Entering the lift lobby of Yilong you are greeted by a speakeasy style concierge window. Leave your coat and take your room key as you are then transported into the world of mid-century high society. Brass accents, walnut timbers, custom terrazzo flooring and distressed vintage rugs adorn this opulent space whilst joinery details have been carefully crafted to mimic a traditional silhouettes. The floorplan positions the main bar in the middle of the tenancy making it two sided with archways that staff can walk through to reach either side. This also divided the space between the brighter area that's more tea focused and the cocktail bar that houses the impressive walk in wine display and is more liquor focused. A big part of the brief and subsequent interior inspiration was that the vertical circulation needed to be grand, much like a hotel lobby of the era. The base footprint of the stair therefore splays out at the end in a sweeping motion, terminating right at the tea bar. The landing in the middle was elongated so patrons can pause and visually take in the entire space, mezzanine and void with a custom designed balustrade and lead light feature pane. Up on level 6 the timber takes on a darker wenge hue and this is where the more premium rooms are housed with two VIP private dining rooms divided by an acoustic bifold door for maximum flexibility.

Project details

▶ **Interior Design:** BrandWorks
▶ **Homepage:** http://www.brandworks.co/
▶ **Branding Design:** BrandWorks
▶ **Area:** 380 m²
▶ **Location:** 20 Dong Hu Road, Xu Hui District, Shanghai, China
▶ **Photographs:** StudioSZ Photo

About us

Advocating design-led thinking, we work to drive a positive impact on people, communities, and the economy so that we may all prosper in the future. We align with business leaders, entrepreneurs and changemakers who believe that their business can do better through design. By understanding their bandwidth, we set them up for growth and scale and meet at the intersection of transformation and success.

BrandWorks are a multi-disciplinary team committed to developing strategic design solutions for FMCGs, Hospitality, Commercial Property, Digital Online brands and Fit-for-Purpose Retail Destinations. We have offices in Melbourne, Newcastle, Jakarta and Changsha, China. What makes us stand out from the crowd is that we've been there. From business management to ownership to operation, our experiences with and love for great design guides our process. We believe that our clients' success supports us and vice versa. So together, we're building a global business community curated for success with innovative and intelligent design.

Project Innovation/Need: The booth areas are upholstered in leather on level 5 and velvet on level 6, and have been finished with contrast velvet piping. This piping detail was inspired by a typical Chinese cheongsam dress that was originally designed in Shanghai in the 1920's. Behind the banquettes the wall panels have been upholstered in cheongsam material and walnut timber joinery mimics the iconic clasp design of the dress. This subtle nod honours fashion houses of the era and provides a joinery language that is repeated throughout the entire venue. The client wanted the lighting to be customisable for customers. In addition to working directly with lighting consultants 'lighting spaces' to design a flexible scheme, we also installed small spotlights above the tables so patrons would be able to take photos of their meals for social media. This light is then dimmed when they begin dining so as to not take away from the mood. Finally, because the space caters to both morning tea through to late night cocktails, we needed to ensure the design was appropriate for any time of the day. By placing the bar in the centre of the venue on level 5 you are able to enjoy it as a focal point wherever you are seated as there is a 360 degree view to the bar from wherever you are sitting.

壹笼
壹笼

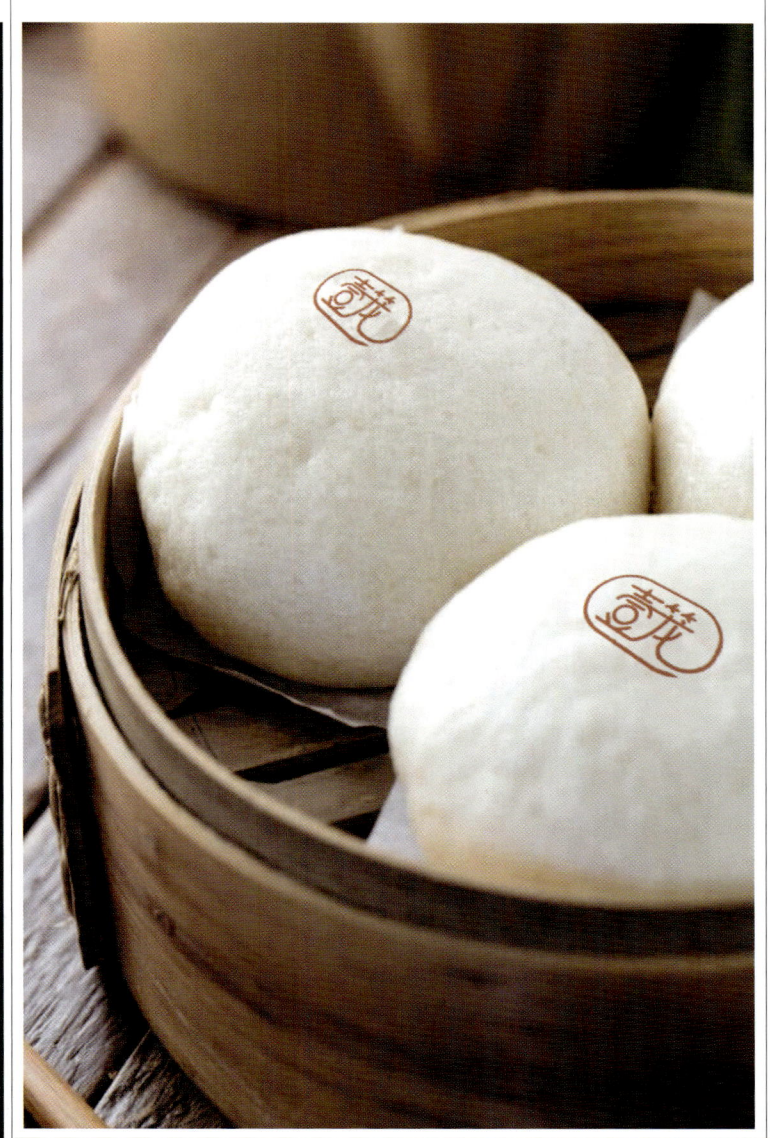

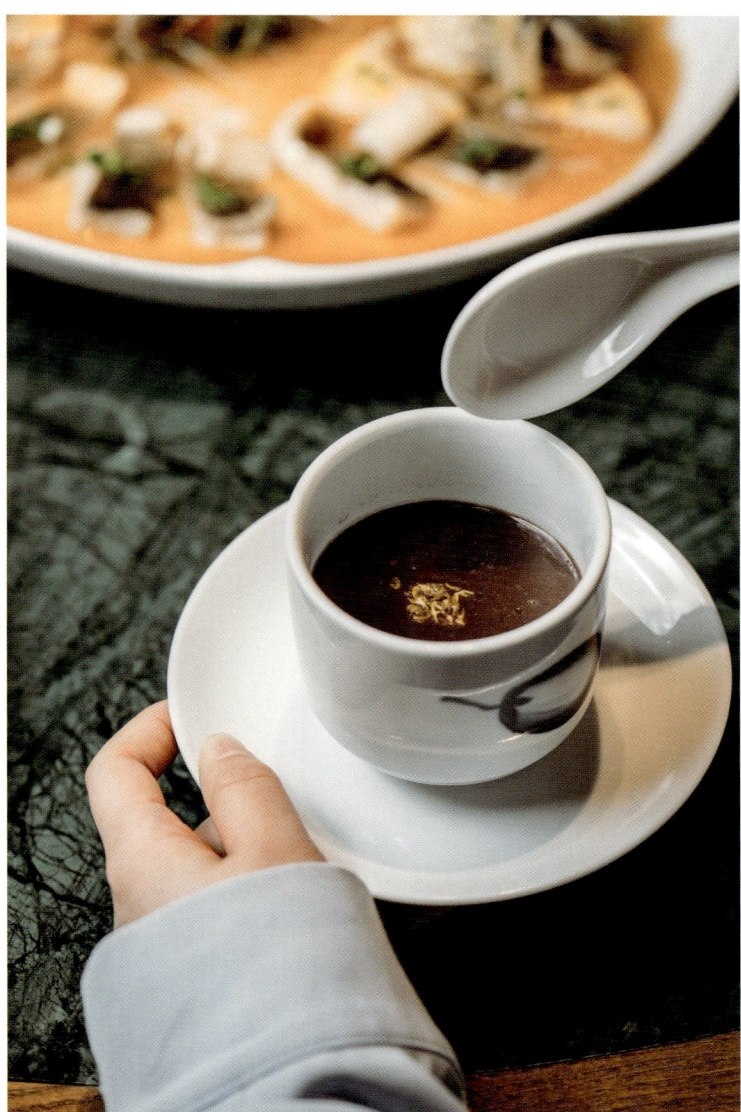

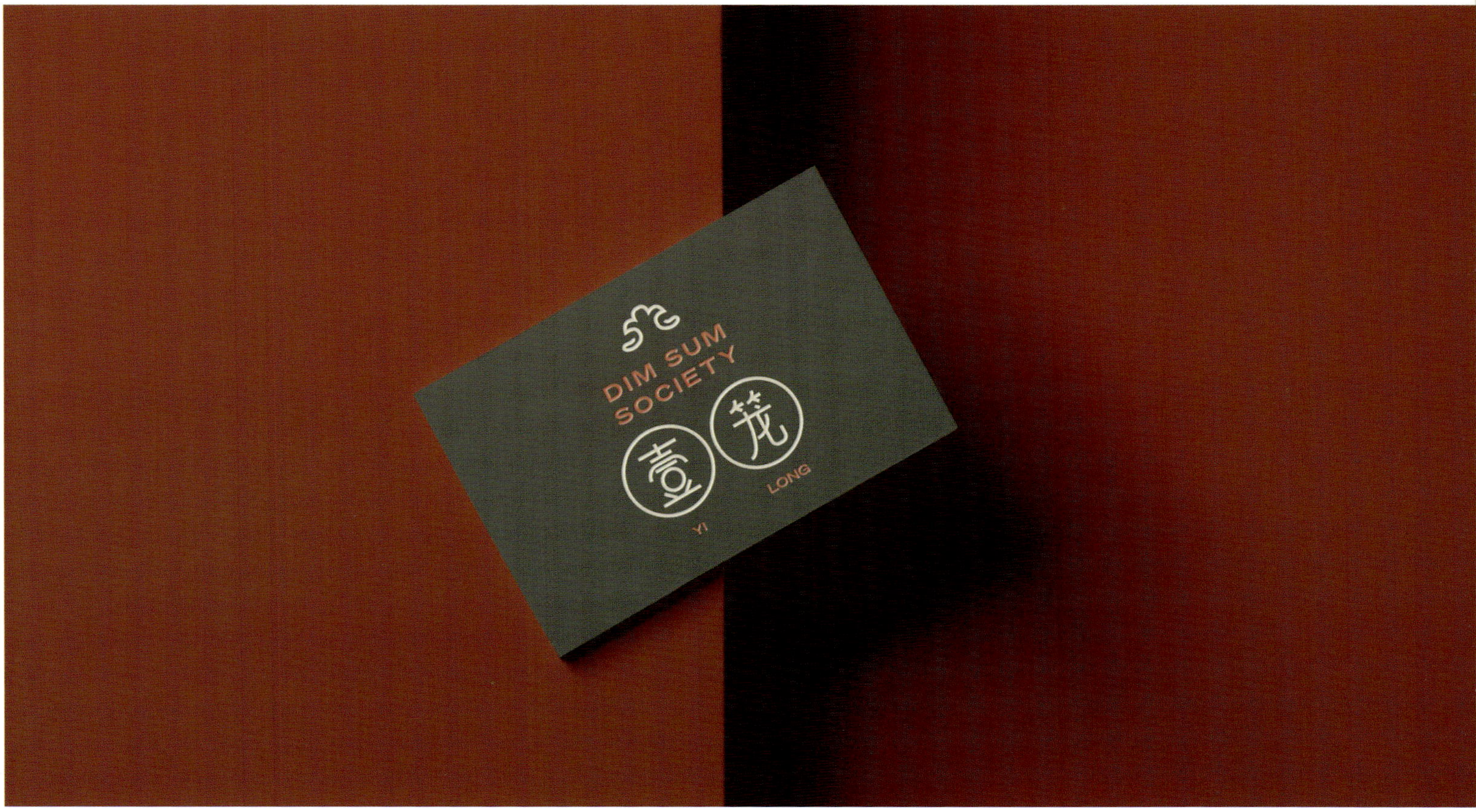
DIM SUM
SOCIETY
壹
笼
YI
LONG

Sustainability: Natural materials were chosen where possible for Yilong with timber, steel, stone, linen and silk being featured. Whilst finishes were chosen in Australia, all items were then locally sourced to avoid the carbon footprint of importation.

壹
笼

点心菜单
饮料
点心菜单

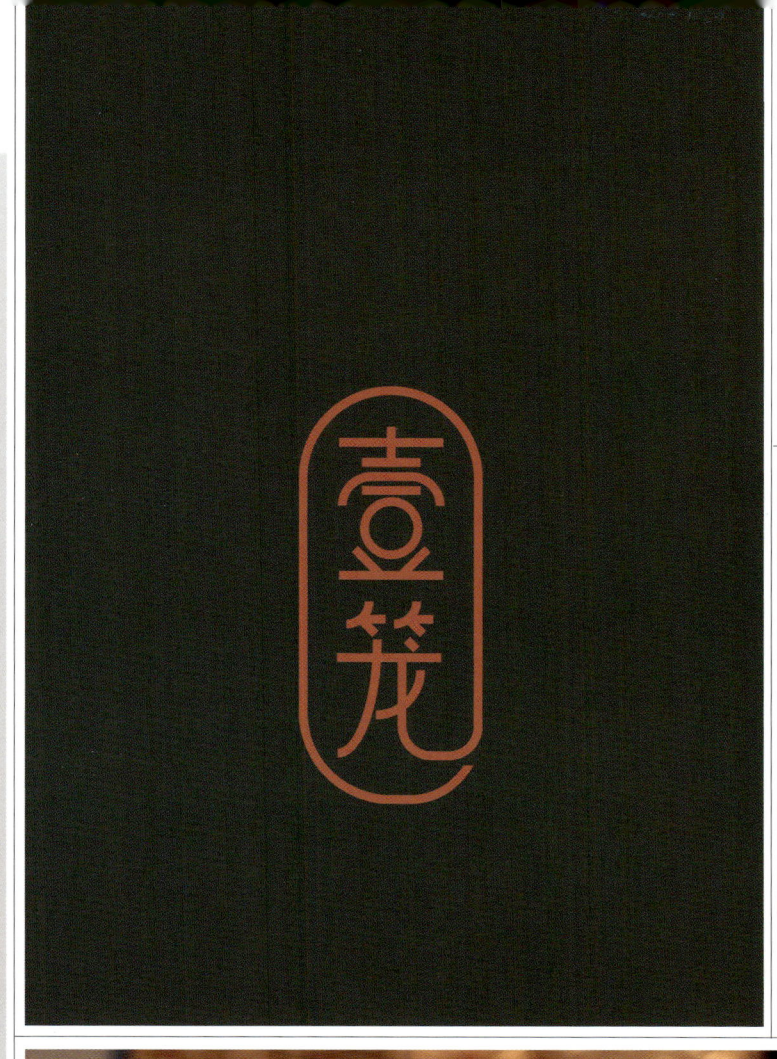

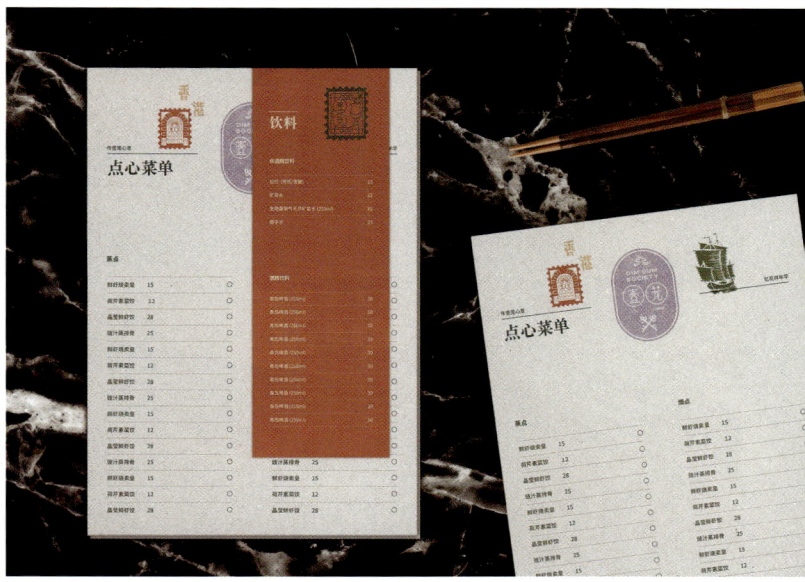

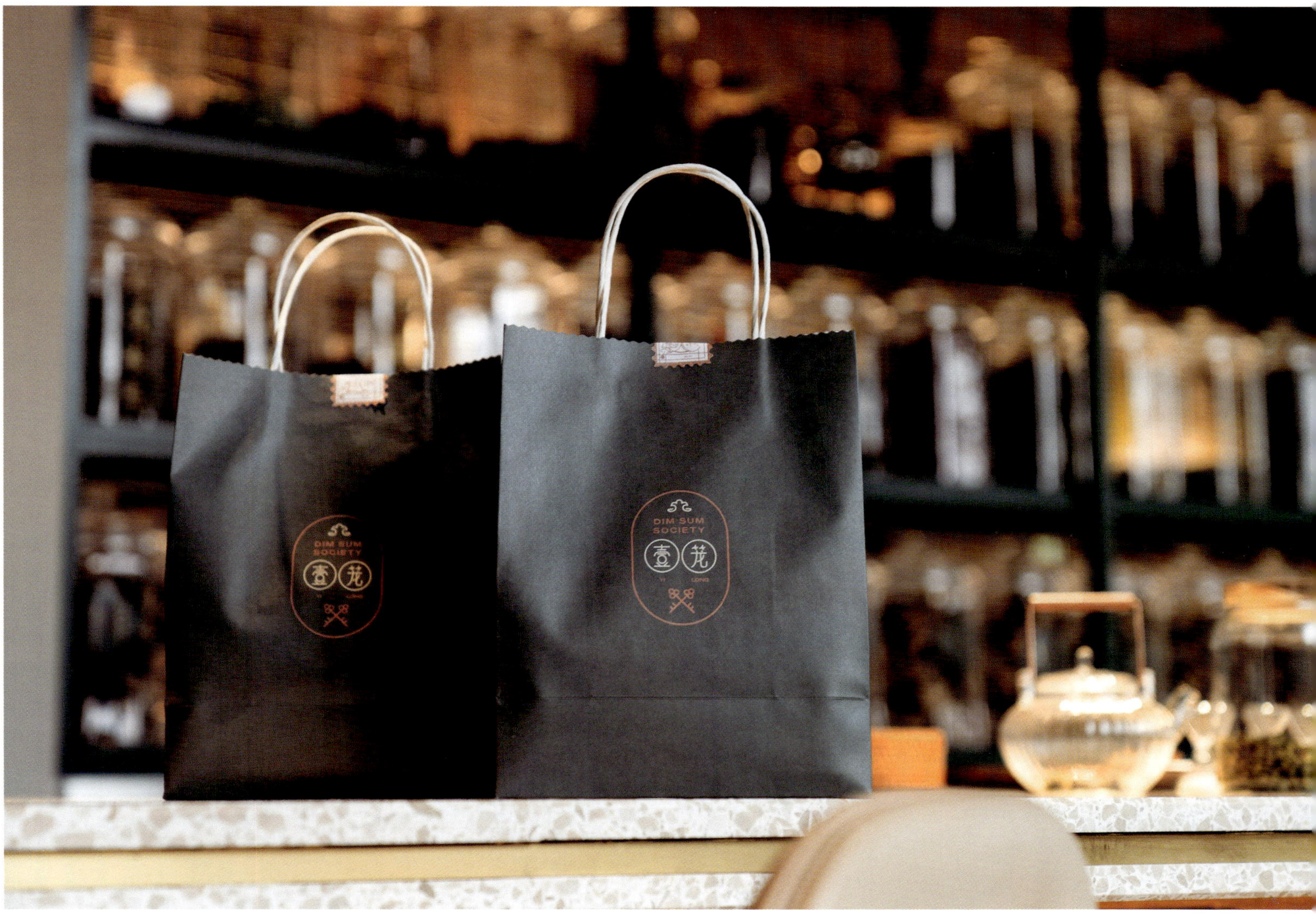
DIM SUM
SOCIETY
壹 笼

Design Challenge: The major challenge in this design was that a lot of design work was executed in Melbourne during the height of the pande All finish samples were chosen in Melbourne during lockdown and then f over for the team in Shanghai to source. The Australian team also ha work directly with our Chinese team to ensure the project was hitting the right tones culturally, so that the venue would appeal to not only l but be a destination venue for travellers in a post Covid world. Choosin cheongsam material was particularly laborious as there are some prints are solely worn for weddings, so there was a lot of back and forth unti perfect print and tone was chosen. Construction Documentation was completed by the Australian team which then needed to be fully translate the construction team over in Shanghai. BrandWorks designed and d mented the large concrete staircase that was then structurally enginee over in Shanghai with minimal disruption to the aesthetic design.

Interior Design: This award celebrates innovative and creative building interiors where people eat and drink - this includes bars, restaurants, cafes and clubs. Judging consideration is given to space creation and planning, furnishings, finishes, aesthetic presentation and functionality. Consideration also given to space allocation, traffic flow, building services, lighting, fixtures, flooring, colours, furnishings and surface finishes.

Filled with nostalgia and defining moments in Hong Kong's pop culture and travel, the story begins with the "Grand Destination", reminiscing the art of travel and the destination as a journey unto itself. The British colonial outpost set in the tropical beach-side scenery of Hong Kong served as a breezy backdrop for the art-deco inspired venue; a leisurely summer afternoon, with the cool sea breeze across your face, providing relief from the humidity and tropical sun.

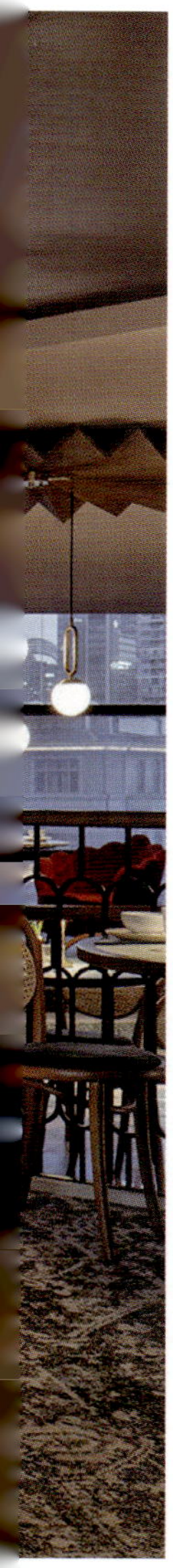
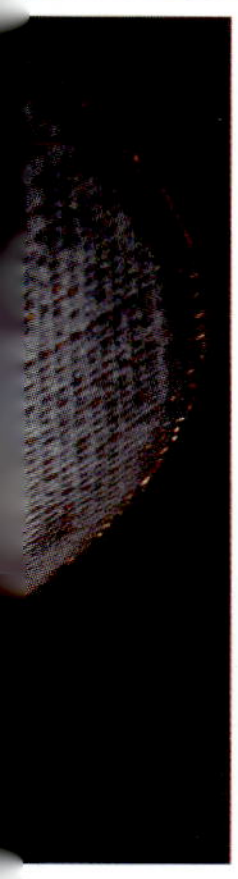
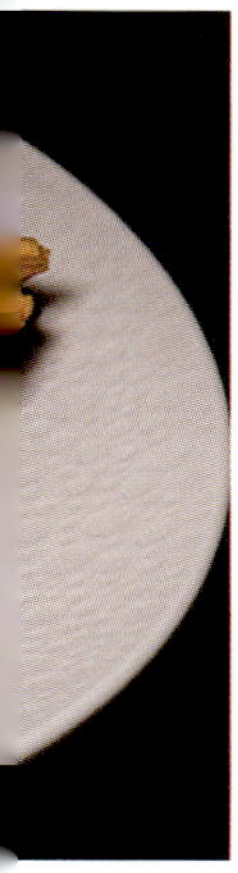
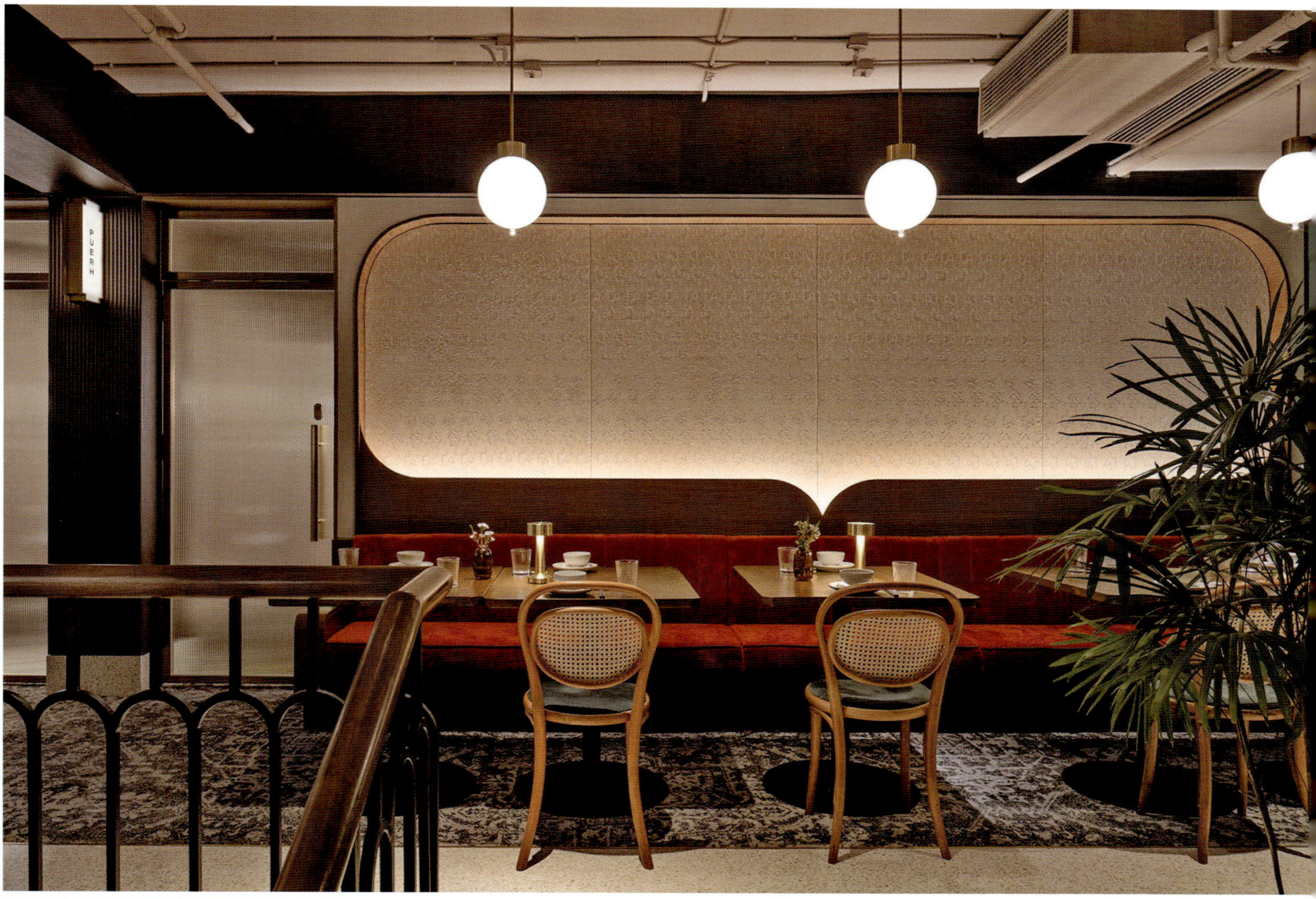
PUERH
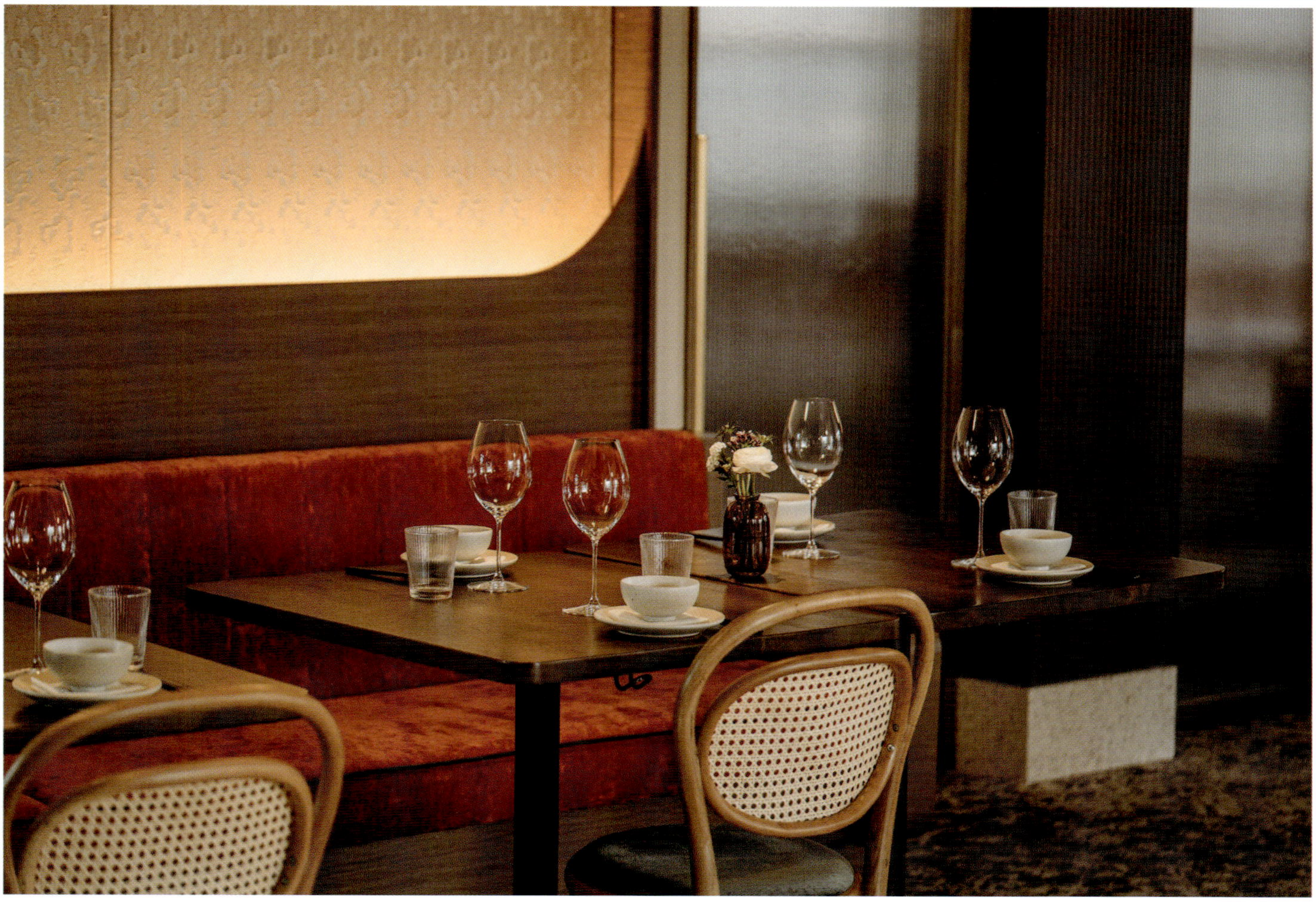

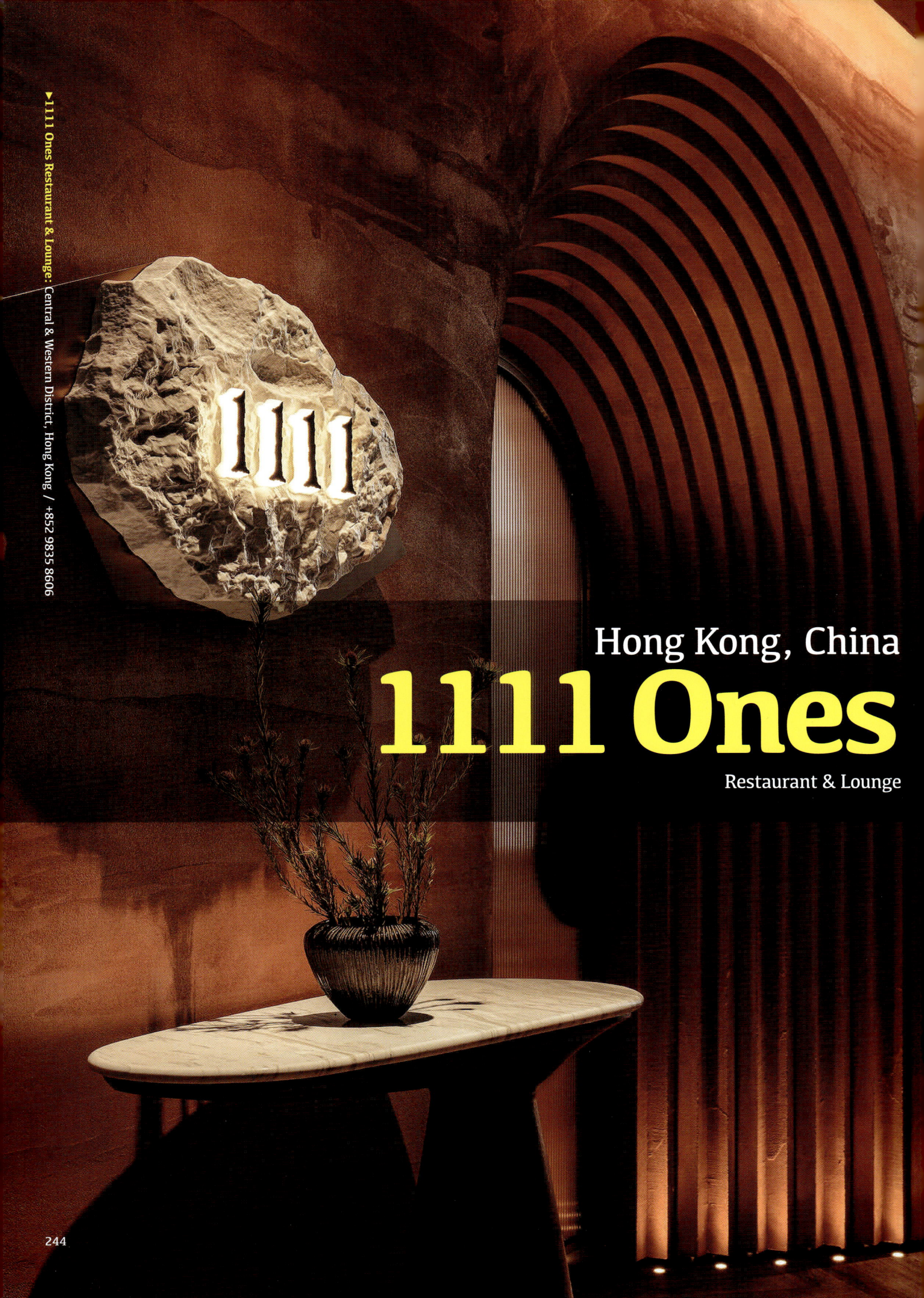

▶ 1111 Ones Restaurant & Lounge: Central & Western District, Hong Kong / +852 9835 8606

1111 Ones
Hong Kong, China
Restaurant & Lounge

The new addition to On Lan Street's fine dining scene – 1111 Ones Restaurant & Lounge brings gastronomes an original collaboration between creative culinary arts and spectacular landscape inspirations. Teamed up with the acclaimed landscape photographer Kelvin Yuen, 1111 Ones curates an enchanting dining experience blending modern Cantonese cuisine with an artistic geological flair. Named after a powerful universal number composed in a repetitive pattern, 1111 Ones connotes new beginnings of all things in life. Designed by Myron Kwan, founder and creative director of local interior design firm M.R. Studio, 1111 Ones draws inspiration from the picturesque Antelope Canyon in Arizona.

Captivated by Antelope Canyon's supernatural appearance, Kwan takes cues from the Canyon's wave-like structure and geological rock formation, transforms the iconic organic forms and shapes, textures and colour palette into a surreal dining space that is unconstrained by conventionalism. Adopted a singular design narrative informed by the natural scenery elements, Kwan effortlessly infuses contemporary flair into the space. The interior results in a placid place, reminiscent of the ravishing and tranquil environs of Antelope Canyon. Upon arrival, the striking burnt orange hued entrance foyer resembles the signature colour palette of Antelope Canyon. The brass base sets the organic-shaped marble plaque off to advantage, exuding a sophisticated sense of beauty. Thoughtfully curated plant arrangement permeates an exotic energy. The symmetric arched doorway is designed with layer upon layer, with the floor lights projecting up, the feature entrance reimagines the famous light beams that shine across the sandstone in the canyon and shape the swirling effects. Radiating an awe-inspiring ambience, the mythical beauty of entrance foyer lure diners to further explore the space and let the journey unfold.

Project details

▶**Interior Design:** M.R. Studio
▶**Homepage:** https://mrstudio.hk
▶**Area:** 186 m²
▶**Location:** Hong Kong, China
▶**Photographs:** Steven Ko

About us

M.R. Studio Ltd. (M.R.S) is a design studio known for its specialization in hospitality, commercial, private residential, and office projects. Founded by award-winning designer Myron Kwan, he creates multi-faceted couture interiors, advising on every aspect and detail of a project through his expertise in space planning, conceptual and design development, and instilled with a heightened sense of narrative.

Our style is, we believe, enthralling, elegant, and timeless. In which, the work of M.R. Studio employs a range of ideas, lines, materials, objects, and references – from unexpected to ever-evolving. We believe every element of our handcrafts environments enable our very appreciative clientele to share our obsession with detail, craft and refinement.

▶**Contacts** e-mail: info@mrstudio.hk / Call us: +852 9747 1278

Entering the restaurant, diners are immediately immersed in an ethereal world formed with undulating curves and fluid forms. Brushed with hand-painted copper hues, the main dining room continues the same colour palette seen from foyer, with the subtle veins and textures organically flowing around the space. The undulating lines of the ceiling feature define the space, capturing the beauty of the distinctive "flowing" shape in the sandstone.

The white marbled countertop kitchen table anchors the main dining room, while its timber upholstery base continues the characteristic "flowing" shapes, echoing the feature curvaceous silhouette. A bespoke pendant lighting following the ceiling curves, its layers of copper compose an artistic impression resonating with Roman number Ones, echoing the restaurant's name. Carefully selected furniture following the fluidity nature, the subdued colour tone and soft curves together cultivate a feeling of ultimate comfort.

The similar craft and design elements, such as symmetric arched door, featured floating pendant lighting, and sophisticated furniture are as well adopted in the private dining room, creating a strong coherent design that enchants its audience like no others.

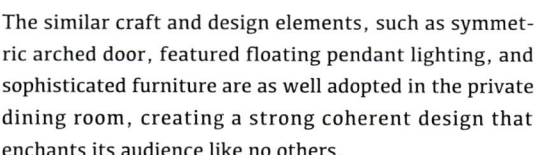

The similar craft and design elements, such as symmetric arched door, featured floating pendant lighting, and sophisticated furniture are as well adopted in the private dining room, creating a strong coherent design that enchants its audience like no others.

Fujiwara Yoshi
Kiev, Ukraine
Japanese Restaurant Fujiwara Yoshi

Japanese Restaurant Fujiwara Yoshi: Solomenskaya St, Kyiv, Ukraine / +380 67 400 7818

Sergey Makhno Architects have designed the largest Japanese restaurant in the Ukraine on request of the chef Fujiwara Yoshihiro. Fujiwara Yoshi is an 800 square meter maze designed to get lost in the greatness of Japanese culture and not even try to find a way out. Take a look at the complete story after the jump. From the architects: "This project is special to me. Japan is my place of power. I fall in love with this country every time I go there. And I always want more. Therefore, I was happy to put a piece of my love in the heart of Ukraine. Our main task was to communicate Japanese philosophy, not to shout it with some clichés. The design of the restaurant is coziness dictated by Japan, but read with the Ukrainian soul" , -says Sergey Makhno, the founder of the Sergey Makhno Architects studio.

The labyrinth stretches past the terrace-garden. Panoramic windows fill it with a flood of daylight. Author's lighting by Makhno floats underneath the black ceiling: the minimalist Gemini, the laconic ceramic Runa, and if you look into the Lakuna lamps, you will see Japan itself. Floor lamps, made in the form of the chasen (a whisk for matcha tea), sow tender light directly to the guests' plates. The garden throws the shadows at the sitting area. Catch one or two, when you are around. The alley of bonsai trees brought from a small town near Tokyo leads to sea-life and freshwater aquariums with a shamelessly bulky table for tuna processing behind.
Turn right.

Project details

▶**Interior Design:** Sergey Makhno Architects
▶**Homepage:** https://mahno.com.ua
▶**Area:** 800 m²
▶**Location:** Kiev, Ukraine
▶**Photographs:** Andrey Avdeenko

About us

Since 2003, we have been creating projects intertwining Ukrainian traditions, world trends and the Japanese philosophy of wabi-sabi, the forces of nature, human talent and architectural mind. We have completed over 600 projects in 21 countries and are ready to create something significant for you. The main task of our specialists — architects, designers, ceramists — is to create comfort that will become near and dear.

▶Contacts e-mail: office@mahno.com.ua / Call us: +38 067 555 55 15

FUJIWARA YOSHI

The sushi bar is already making instant rolls under large Fuji-lamps. Sergey Makhno personally designed them, contemplating that snow-white Japanese mountain. Instead of the ordinary lighting, there are chimneys in one tatami-room, and a plump copper flower in another. They can tell you about the sun hiding behind the roofs in the Japanese gardens. For business negotiations, there is a separate room with wooden walls, floor, and ceiling-so that nobody overhears. And in a large banquet hall, under the poppy-heads of lamps, a table is threading across the room.

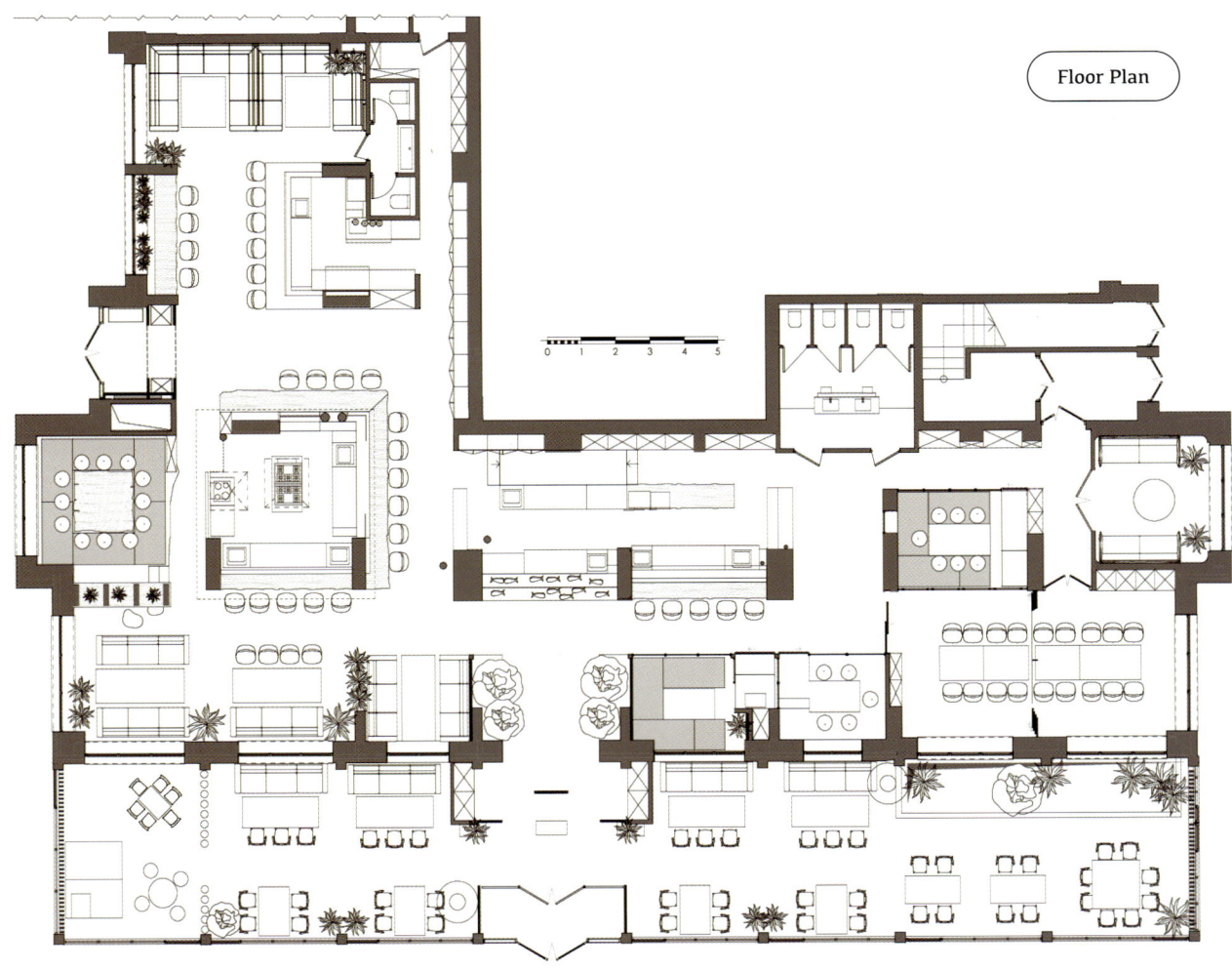

Floor Plan

■ Under the bamboo chasen sprouts, soft-cream seats and chairs embosom the tables. Everything is made specifically for this restaurant. Next to them, the tempura and robata grill zones are located. In the tatami-room #3, the sky stretches out on the wall and the chairs, while ikebana tickle guests with flowers-branches. The ceramic barrels of sake and a mob of the wine refrigerator prompt what will go farther. Farther, will be something winy — under a light graded cloth, in the arms of the travertine, the bar is ringing with glasses. Just in time, because the stone slabs on the wall begin to tell "One hundred poems of one hundred poets".

"Most Japanese restaurants are low-key, small and definitely cozy. Therefore, we paid a lot of attention to the layout in order to host 250 guests at the same time. And yet, there is an atmosphere of coziness and peace of mind, says Illia Tovstonog, the chief architect of Sergey Makhno Architects. Fujiwara Yoshi is a Japanese restaurant with Ukrainian soul. It is a maze you don't want to leave. Here, aesthetics waltzes with functionality, pipes become lamps, flowers grow from the ceiling, and moss - from the plates. It is a place to rest your eyes, hold your breath and savor Japan.

The sushi bar is already making instant rolls under large Fuji-lamps. Sergey Makhno personally designed them, contemplating that snow-white Japanese mountain. Instead of the ordinary lighting, there are chimneys in one tatami-room, and a plump copper flower in another. They can tell you about the sun hiding behind the roofs in the Japanese gardens. For business negotiations, there is a separate room with wooden walls, floor, and ceiling — so that nobody overhears. And in a large banquet hall, under the poppy-heads of lamps, a table is threading across the room.

Under the bamboo chasen sprouts, soft-cream seats and chairs embosom the tables. Everything is made specifically for this restaurant. Next to them, the tempura and robata grill zones are located. In the tatami-room #3, the sky stretches out on the wall and the chairs, while ikebana tickle guests with flowers-branches. The ceramic barrels of sake and a mob of the wine refrigerator prompt what will go farther. Farther, will be something winy - under a light graded cloth, in the arms of the travertine, the bar is ringing with glasses. Just in time, because the stone slabs on the wall begin to tell "One hundred poems of one hundred poets".

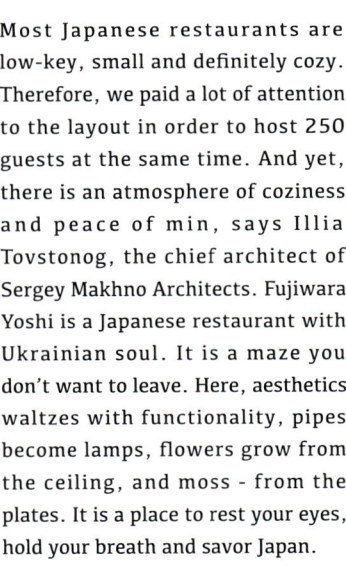

Most Japanese restaurants are low-key, small and definitely cozy. Therefore, we paid a lot of attention to the layout in order to host 250 guests at the same time. And yet, there is an atmosphere of coziness and peace of min, says Illia Tovstonog, the chief architect of Sergey Makhno Architects. Fujiwara Yoshi is a Japanese restaurant with Ukrainian soul. It is a maze you don't want to leave. Here, aesthetics waltzes with functionality, pipes become lamps, flowers grow from the ceiling, and moss - from the plates. It is a place to rest your eyes, hold your breath and savor Japan.

Here, aesthetics waltzes with functionality, pipes become lamps, flowers grow from the ceiling, and moss - from the plates. It is a place to rest your eyes, hold your breath and savor Japan.

Margo restaurant

Hong Kong, China

Brasserie and Bar Meet at MARGO and Kyle & Bain - Japanese & Spanish restaurant

Designed by Myron Kwan of M.R. Studio, MARGO and Kyle & Bain come as an elegant pair of brasserie-style eatery and martini bar that nods to elements of the city's history

MARGO's brightly lit glass facade on Central's Ice House street is a promising first impression of the new Hong Kong brasserie-and-bar duo inside. Spanning 230 square metres over two levels are MARGO, a modern European brasserie and, on the restaurant's mezzanine floor, martini bar Kyle & Bain. The two spaces are distinct, yet naturally coherent thanks to their smart design. 'The concept of bringing two F&B experiences into one location actually came from the client. He wanted guests to be able to enjoy a drink right after having a good meal,' says Myron Kwan, founder of M.R. Studio, who undertook the project for hospitality group Leading Nation. 'Our main intention was thus to create a connection between the brasserie and the bar.'

On the ground floor, MARGO's plush, pink interiors are characterised by plenty of timber, with pink velvet banquettes, terrazzo table tops and plant installations adding lively details. Those acquainted with Leading Nation's popular Hong Kong cafe brand Elephant Grounds will recognise similarities between the two — an intentional move to create a sense of familiarity for customers.

Project details

- **Interior Design:** M.R. Studio
- **Homepage:** https://mrstudio.hk
- **Area:** 230 m²
- **Location:** Queen's Road Central, Hong Kong, China
- **Photographs:** M.R. Studio

About us

M.R. Studio Ltd. (M.R.S) is a design studio known for its specialization in hospitality, commercial, private residential, and office projects. Founded by award-winning designer Myron Kwan, he creates multi-faceted couture interiors, advising on every aspect and detail of a project through his expertise in space planning, conceptual and design development, and instilled with a heightened sense of narrative.

Our style is, we believe, enthralling, elegant, and timeless. In which, the work of M.R. Studio employs a range of ideas, lines, materials, objects, and references – from unexpected to ever-evolving. We believe every element of our handcrafts environments enable our very appreciative clientele to share our obsession with detail, craft and refinement.

▶ Contacts e-mail: info@mrstudio.hk / Call us: +852 9747 1278

Up a wooden staircase, MARGO's bright and casual setting gives way to Kyle & Bain's more intimate and subdued atmosphere. Taking its name from Scottish engineers William Kyle and John Bain who established Hong Kong's first ice plant in the 1870s on the building's very site, the martini bar was designed with this reference in mind. 'We wanted to create an iconic space that's related to the history of the location,' Kwan explains. 'So the bespoke bubble glass lights, for example, are inspired by ice cubes - the final product of the ice plant.'

While the wooden elements create subtle continuity with MARGO, Kyle & Bain has a more muted, lavish feel. 'We chose dark oak wood for the bar, because we wanted that darker, moody atmosphere,' the designer says. Creating a curved canopy above the bar are large symmetrical mirrors that resemble shattered ice, while below, custom velvet bar stools and banquettes and dark marble tables instill a sense of sophistication.

MARGO

Drawing inspiration from the location's history and from his client's existing projects, Kwan has created a dual space that keeps its first promise: a modern, refined and elegant experience that calls for a second drink.

Margo is an intimate brasserie-style European restaurant from Elephant Grounds, the group behind Spanish restaurant La Rambla by Catalunya, Japanese restaurant Wagyumafia, and cafe Elephant Grounds. The designer fashioned the dining room atmosphere with different tones and textures that established an intimate yet Instagrammable feel. The seating plan also encourages intimate conversations of guests.

Plush, pink interiors dominate the interiors at Margo, a modern European brasserie that is also home to the martini
Kyle & Bain. Everything here seems to be designed for maximal visual impact: created by Hong Kong-based inter
design studio MR Studio, it features coral pink velvet banquettes; an abundance of greenery, including an oversized m
wall that acts as the centrepiece of the wood-dominated space; and details like terrazzo tabletops. Plenty of thought a
went into the presentation of the dishes. An octopus terrine, served in two ethereal bite-sized cubes, resembles
terrazzo of the dining tables, while a beef tartare has a flashy look not unlike a Jackson Pollock painting.

Restaurant Lunar

Shanghai, China

Lunar is a Modern Chinese restaurant situated inside a standalone villa at the heart of Shanghai on West Jianguo Road

So Studio creates the space in the hope of conveying a sense of tranquility and comfort, drastically different from the bustling and dazzling city chores. The interior design is not only focused on the Chinese traditional structure details but also pays attention to the big picture of the beauty aesthetic. Coheres the ambiance with oriental design basics to an appropriate scale, integrating geometry ratio. Lunar is a Modern Chinese restaurant situated inside a standalone villa at the heart of Shanghai on West Jianguo Road, which is running by the opposite group and chef Johnston Teo. Termed Modern Chinese, the cuisine at Lunar is to celebrate the quintessential way of Chinese dining. The namesake pays homage to the Chinese lunar calendar. Resonating with the lunar customs is essential to the ethos of dining experience design. Embracing the seasonality and terroir is at its core.

The tea stall located on the first floor. The soft lighting showered over the shoulder, conveying a sense of comfort, combines with the scent of tea brick and bamboo curtain immerses one in a moonlight night the moment of stepping in. Ascending along the stairs, the passing pebble-paved trail would land one on the second floor of Lunar. A unique hanging cornice, made possible by the outlining curves of classical Jiangnan garden fusing into the architectural structure of the space, guides the way connecting to the dining area and private dining room. The mural hand painting renders orange fluorescence like moonlight shadow, echoing the shooting tea stall on the first floor. Collaborating with the florist and designer Maggie Mao to add an orange haze to space, the small and delicate decoration adds a touch of Zen in the simplicity. The most important material in the whole space is the matte surface. Without excessive reflection, conveying calm and beauty through the texture and touch.

Project details

▶**Interior Design:** So Studio
▶**Homepage:** https://www.sooostudio.com
▶**Area:** 260 m²
▶**Location:** Shanghai, China
▶**Photographs:** Wen Studio

About us

Founded in 2016 by partners Yifan Wu and Mengjie Liu, Sò Studio is an innovative interior, and industrial design practice based in Shanghai. Its scope of work spans from commercial complex, retail space, hospitality, public institution, office, medical space, and private residential. Sò Studio is composed of a dynamic team with diverse background, of whom have practiced in art and design with a global worldview.

Sò Studio conducts a comprehensive research prior to schematic design, and we work closely with our clients to employ a bespoke design approach by following the authenticity of each project with emphasis on the cultural influence and unique form of the site. Our projects range from varied scales and types, from creative concept for retail stores, hospitality to collaborating with top developers to re-image commercial interiors, hence, the diverse explorations enable us to place an emphasize on expressing narrative in spatial design. By virtue of applying imagination in our design language to create inspiring space with powerful visual impact, Sò Studio aspires to deliver a better vision of life and probe into the society values while curating spatial design and user experience.

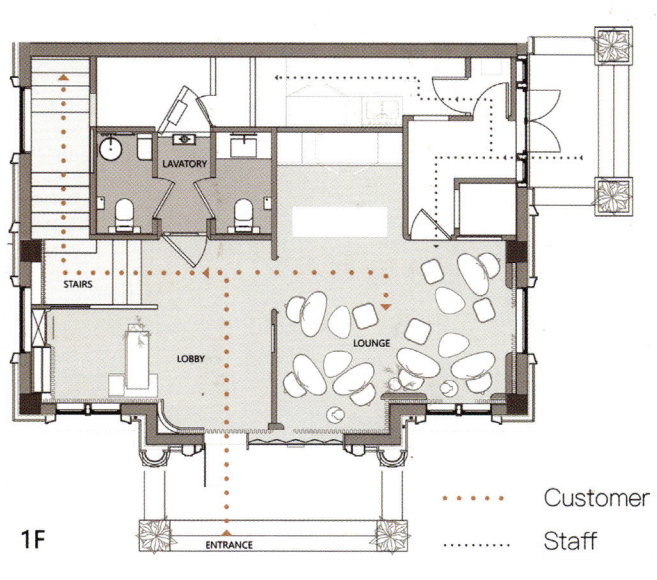

1F

· · · · · Customer
· · · · · · Staff

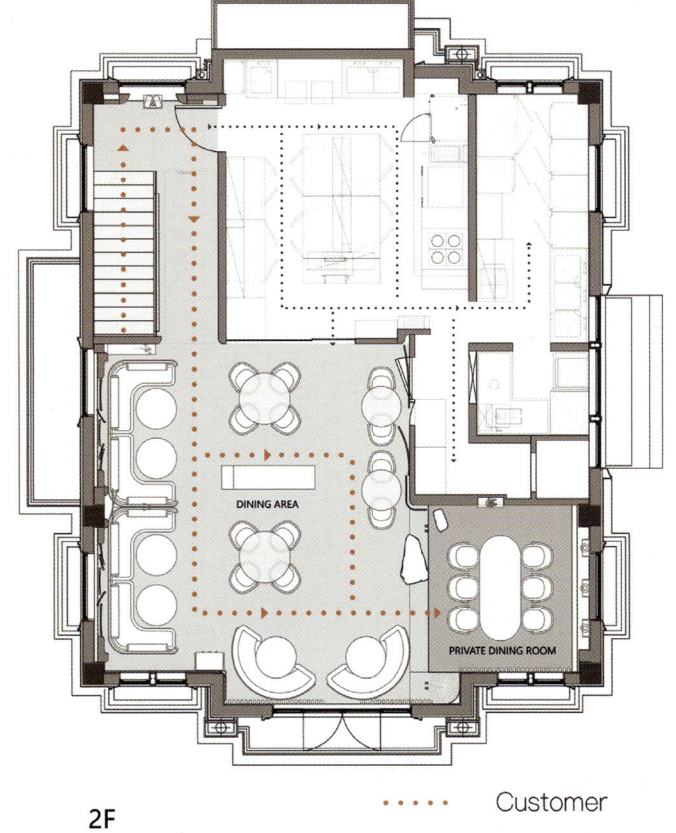

2F

· · · · · Customer
· · · · · · Staff

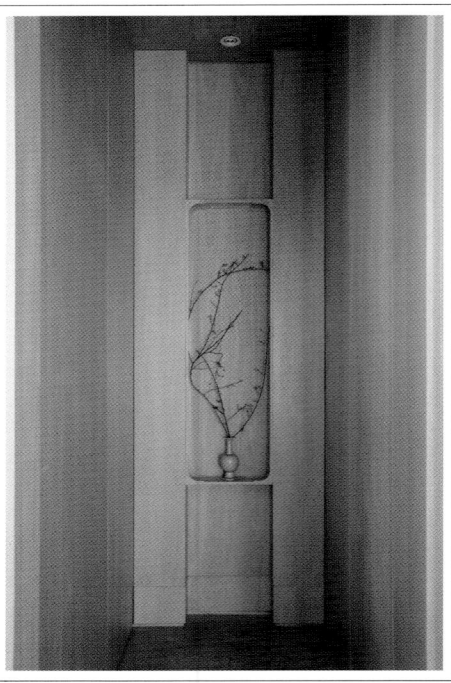